# What Every Engineer Should Know About Product Liability

# WHAT EVERY ENGINEER SHOULD KNOW
A Series

*Editor*

**William H. Middendorf**

*Department of Electrical and Computer Engineering*
*University of Cincinnati*
*Cincinnati, Ohio*

**Vol. 1** What Every Engineer Should Know About Patents, *William G. Konold, Bruce Tittel, Donald F. Frei, and David S. Stallard*

**Vol. 2** What Every Engineer Should Know About Product Liability, *James F. Thorpe and William H. Middendorf*

*In preparation*

What Every Engineer Should Know About Microcomputers, Part 1, Hardware/ Software Design: A Step-by-Step Example, *William Bennett and Carl Evert*

*Other volumes in preparation*

# What Every Engineer Should Know About Product Liability

James F. Thorpe
William H. Middendorf

*College of Engineering*
*University of Cincinnati*
*Cincinnati, Ohio*



**MARCEL DEKKER, INC.** **New York and Basel**

**Library of Congress Cataloging in Publication Data**

Thorpe, James F [Date]
What every engineer should know about product liability.

(What every engineer should know ; v. 2)
Includes bibliographical references and index.
1. Products liability--United States. 2. Engineers--
United States--Handbooks, manuals, etc. I. Middendorf,
William H., joint author. II. Title.
KF1296.T48 346'.73'038 79-15963
ISBN 0-8247-6876-0



Marcel Dekker, Inc.
270 Madison Avenue, New York, New York 10016

Current printing (last digit):
10 9 8 7 6 5 4 3 2 1

Printed in the United States of America

# Preface

This book evolved from work done by one of the authors during a sabbatical leave several years ago. At that time, the intention was to prepare a monograph summarizing the research done during the sabbatical in the area of product liability. It was intended that the monograph would be rather theoretical and have a limited distribution.

During the past two years, it has become increasingly evident that a more practical book is needed, one that is directed toward the needs of the design engineer and that would also be suitable as a reference book for engineering students taking design courses. The need for such a book is evidenced by the recognition of professional engineering societies and a few engineering educators that legal problems are beginning to have a great impact upon the engineering profession. For example, the American Society of Mechanical Engineers (ASME) has formed a new committee—Design Engineering and the Law—in order to provide a forum for discussion and scholarly work in this area.

Because the editor of this series was also engaged in many facets of product liability, a joint authorship evolved quite naturally in which the experiences of both a mechanical and an electrical engineer could be synergistically combined. The authors have both been involved in product design, in interpreting standards, and in litigation as expert witnesses for both plaintiff and defendants in product liability cases. Nevertheless, we are primarily engineering educators and not attorneys.

We have tried to write a book that will prove useful primarily to the engineer who is (or may become) involved in product design, product safety, product management, or product litigation. The process of designing safer products does not require new aptitudes or abilities not already possessed by engineers who find themselves buffeted by the many pressures of total design responsibility. What is required is a broadened perspective and recognition that the process of designing safer products (and lessening product liability risk) involves a natural extension of the traditional functional design procedures to include the additional considerations outlined in this book. We have attempted to collect, organize, and present these topics in a reasonably self-contained and sequential development that will enhance the continuing education and stimulate the latent desire of the design engineer to fulfill his total responsibility as a designer. We have also provided advice for the neophyte engineering expert witness and discussed some of the troublesome ethics problems that may confront the engineer as he weighs loyalty to his employer and his professional responsibility to protect the public. And, as engineering educators, we have had something to say about the role of product liability in engineering education and what it portends for the future of design education.

We hope that this book will also prove useful to lawyers, insurance professionals, business persons, and faculty members as well as to engineering students taking design courses. In addition to information provided as a result of our own personal experiences in product liability cases, we have gleaned much more information from the writings and statements of others. Our contribution has been to synthesize and organize this material. In doing so, we want to emphasize that we do not wish to be construed as advocates for any individual party having a vested interest in product liability litigation. We can see merit in the position of the injured party and in the position of the product manufacturer. The insurance industry, the legal profession, the government, and the engineering profession also have legitimate interests.

We have tried to make clear that this subject is complex and undergoes continuous change. To the best of our ability we have accurately stated the law and the appropriate engineering counsel. However, we must emphatically state that the law does differ from court to court in the United States. Further, the law and its practice is responsive to each new court decision and to new statutes enacted by Congress. Whenever questions arise concerning product liability we strongly recommend that a lawyer specializing in product liability be consulted. The purpose of writing this book for the engineer is not to replace competent legal counsel but to make possible a more meaningful dialogue with him. Therefore we recommend such a dialogue even when the engineer is confronted with a situation quite similar to one described in this book.

The current crises in technology, of which product liability is just one, create new and attractive opportunities for engineers to broaden their influence and to emerge as more complete professionals. This is the direction in which we would like to see both engineering and engineering education move and we hope that this book will be a contribution toward that goal.

Finally, the use of the male gender in the language of the law has been carried over into this book. Hopefully, it is not our male chauvinistic tendencies that has made it seem inconvenient to use what still seems awkward to us, the construction s/he. If our usage offends anyone, we apologize to that person.

# Acknowledgments

The authors wish to acknowledge the use of helpful information contained in numerous publications on product liability. Also, several people have been helpful either in reviewing written material or in providing information. We thank Judge Rupert A. Doan for reviewing our material on the legal system and product liability law. We thank Tom Geile and Karl Krug for help on the insurance materials.

Engineering educators from whom we have learned about product liability include Leo Peters, Reggie Vachon, Tom Talbot, and Al Weinstein. Dr. Weinstein has been a very influential force for bringing product liability into the consciousness of engineering educators and in the preparation of scholarly treatises on the subject.

Finally, we appreciate the comments provided by a group of electrical engineering seniors who reviewed portions of the manuscript.

# Contents

# About the Authors

The authors are professors in the College of Engineering at the University of Cincinnati and engineering consultants in design, product safety, and product liability.

James F. Thorpe is a mechanical engineering graduate of the University of Cincinnati and received his PhD from the University of Pittsburgh. He has direct design experience ranging from conveyor engineering and machine design for a small company to research and analysis in nuclear engineering for a major U.S. corporation. Dr. Thorpe is the author of over forty technical publications, has served as consultant to some forty companies, and serves on several professional society committees concerned with engineering ethical and legal problems.

William H. Middendorf is an electrical engineering graduate of the University of Virginia. He holds an MS from the University of Cincinnati and a PhD from Ohio State. He has twenty-seven patents and has written two prize-winning papers on design. He is a member of the General Engineering Committee of the Low Voltage Distribution Section of NEMA and a member of three Underwriters Laboratories Industry Advisory Councils. Dr. Middendorf is a Fellow of IEEE and received the Herman Schneider award as Distinguished Engineer in 1978.

# What Every Engineer Should Know About Product Liability

# 1

# Introduction

For many years the engineering profession was relatively shielded from legal challenges and profound ethical questions. This comfortable, provincial era no longer exists.

The technologies of nuclear energy, computers, space exploration, microelectronics, and biochemistry have led to a confrontation of the engineering community with the public welfare and social issues such as energy, environmentalism, consumerism, and product safety litigation. Whereas the engineer was once faced with problems of relatively limited scope, the new technologies pose dilemmas that are no longer localized. Consider, for example, that a major accident involving a nuclear power plant may result in effects lingering for many years, and these effects may be irreversible or of global extent.

Thus the subject of this book, product liability, must be viewed in the broader context of an increasing interaction between engineering and society. On the one hand, Congress, the courts, and various consumer agencies are actively involved in establishing legal precedents in order to meet consumer expectations. And, on the other hand, these societal pressures present potentially staggering financial losses for U.S. businesses. The engineer, in the middle of this controversy, has a pivotal role to play.

Traditionally, the interaction between society and U.S. business was characterized by a spirit of *caveat emptor* (let the buyer beware). This same atmosphere permeated design engineering and engineering education, where highest priority was given to strictly technical considerations such as strength of materials, machine elements, electrical components, manufacturability, marketability, and cost. For many years this philosophy remained unchallenged, because the prevailing concepts of liability were based on manufacturer's negligence, contributory negligence of the plaintiff, and reasonable conduct of the manufacturer.

A new era has begun in which business and technology can no longer afford such a narrow vista. Consumer demands and sentiment have been backed by the legal system in moving from concepts of negligence to defect and strict liability. The relationship between manufacturer and user and the design atmosphere have changed to the spirit of *caveat venditor* (let the seller beware). Today the designer must not only be able to apply the technical elements of design but must also meet society's expectation of a design. During the design process itself this means using far broader design criteria, including failure analysis, fail-safe design and redundancy, product safety audits, hazards analysis, and compliance with government and voluntary standards. In managing the design process this means an expanded array of responsibilities, including design review, product liability prevention, and documentation.

There is a legitimate concern on the part of industry and engineering that consumer legislation foreshadows governmental intervention in the design function—legislation that will limit or even usurp the engineer's judgment. Furthermore, overreaction by well-meaning consumer advocates can undoubtedly stifle incentives for innovation and entrepreneurship.

How reasonable are present-day consumer demands? A brief look at product safety statistics supports the demand of consumers for protection from the products of industry. There are some 20 million Americans injured each year as a result of incidents connected with products. Of these, approximately 30,000 are killed and 100,000 are disabled permanently. Even if carelessness by the user is partly responsible, these are shocking statistics.

It is the engineer, in this milieu, who has a unique opportunity to play a professional role. Whether prospectively in product design, retrospectively in litigation, or as an educator training the next generation of professionals, the major responsibilities for the technological decisions that can provide reasonable solutions to the demands for product safety are those of the engineer.

Thus a whole new and broadened perspective opens up to the engineer, which includes issues of technology and society, engineering and the law, ethics, and professional practice. In responding to this challenge the engineer will have to have the interest and will to engage in self-education beyond the traditional areas of his undergraduate education.

# 2

# The Legal System of the United States

The engineer who is involved in product liability has a particular need for a greater appreciation and understanding of the legal system of the United States.

The law and the legal system of the United States evolved from the British legal system, and the legal rights of the American people and the laws governing the nation have their origin in what is called the common law system of England. This is a system of law deriving its force and authority from universal consent, custom, and long usage rather than from express statute. Today, many aspects of this system have been codified.

The training of the engineer is liable to make him prone to view the law simply as a technical institution. This is too narrow a view. Instead, the law must be viewed as a changing, pragmatic institution responsive to the current values and morals of society. The law and the legal system are continuously being questioned and reworked as they attempt to keep pace with the needs of society. The goals of law are justice and fairness rather than the precision sought in science and engineering.

The making and administering of laws in the United States is performed by four major branches—the legislative, the judicial, the executive, and the administrative.

The legislative branch of government has responsibility for the enactment of *substantive law* (statutes or changes in statutes). These are laws that create, define, and regulate rights and duties between citizens and the state and among the citizens themselves.

The judicial branch of government is responsible for *procedural law.* It consists of the machinery that prescribes the methods and means of enforcing substantive law (courts, judges, etc.) and of administrating justice.

The executive branch issues orders and regulations at both state and federal levels, while the administrative branch has broad regulatory powers over industries and activities such as insurance, utilities, transportation, communications, environment, and labor practices. These agencies are also given rule-making authority. In the twentieth century there has been a virtual explosion in the extension of administrative activities and powers.

In addition to its administrative classification, law can be divided into subcategories, all of which are derived from the authority of the Constitution: case law, statutory law, treaties, administrative regulations, and local ordinances. Case law, in particular, is very important in the field of product liability. This is the development of law by the courts as they render decisions and opinions in controversies brought before them.

Obviously, the United States Constitution is the supreme law, and all other local, state, and federal laws must conform to the constitutional provisions.

### Systems of Law

There are two major systems of law in use by the nations of the world. These are the *civil law* system and the *common law* system.

#### *Civil Law System*

The civil law system is followed by most of the non-English-speaking world (excluding the Soviet Union and iron curtain countries). This system of law is derived from Roman traditions and is one in which primary emphasis is on constitutions and statutes as sources of law. In the civil law system *jurisprudence* plays a major role. This is a term used to describe the attempts of legal scholars to provide a theoretical basis for law; jurisprudence thus provides a rationalization and philosophy of law in the civil law system.

In the civil law system, the courts have no power to declare statutes unconstitutional. The statutes are gathered into great codes covering all types of legal matters. In cases in which these codes need interpretation, the opinion of respected jurisprudents is probably more influential than that of judges. In the civil law system, cases coming before the courts do not contribute to the development or become a source of law in the sense of prior judicial decisions having an influence on future decisions.

#### *Common Law System*

The English-speaking world (with the exception of Scotland and a few other states) follows a system of law known as common law. This system of law has its origin in England and was based on the custom of the people. As time went on, the accumulation of cases decided by the courts created precedents which the courts then followed (that is, prior decision = *stare decisis*). Thus, what was originally a system of law based on the custom of the people became a system based on the custom of the courts (case law).

The large body of decisions and judgments found in past decided cases is a major source of law in the common law system. Much of this law is nonstatutory in nature.

This is not to say that statutes do not constitute a primary source of the law in the common law system. However, even with statutory law, the statutes mean only what the courts interpret them to mean. Thus, there is a hierarchy of sources of law in the common law system—namely, constitutions, statues, and cases.

### Restatements of the Law

In the European civil law countries, the concept of doctrinal development (by so-called jurisprudence) is very important. Primary emphasis is given to constitutions and statutes as sources of law.

Even in the common law countries, and particularly in the United States, it is necessary for doctrinal expression to be provided to the law by other means than just the courts. The most general and conscious effort to provide such doctrinal development and to distill the essence from the thousands of court decisions was made by the American Law Institute beginning in the 1930s. At that time, the best legal minds of the nation synthesized thousands of cases and areas of the law into volumes named *Restatements of the Law.*

Even though these restatements are not statutory in nature, they are extremely persuasive in the courts. The doctrine of strict liability, which is the central theme of this book, is such a restatement of the law, known as the *Restatement of Torts Second* (Section 402A). The provisions of this statement will be discussed in detail later.

### Civil Law and Criminal Law

The expression *civil law* is also used in another and completely different context from the one discussed above; it is used to distinguish between criminal and noncriminal law.

In *criminal law* a criminal action is brought by the government against an individual who has allegedly committed a crime. Crimes are classified as acts of treason, felonies, or misdeameanors. Since a crime is an act against society, the criminal court may punish the guilty individuals through fines or imprisonment.

In *civil law* a civil action is brought by a person (an individual, a corporation, or the government) to enforce his private rights. Ordinarily a civil suit involves a dispute between persons involving a breach of agreement or duty imposed by law. In a civil case, the outcome will be either monetary damages or a court order.

This book deals only with civil law and particularly as it relates to the problem of product liability.

**The Judicial System**

The United States has a pluralistic court system consisting of 51 sovereign systems—the federal court system and 50 state systems. The interplay among these court systems regarding questions of jurisdiction can present a very complicated picture.

The U.S. Constitution is the basic document that defines and divides the powers between the federal government and the states. The Tenth Amendment to the Constitution specifies that the "powers not delegated to the United States by the Constitution . . . are reserved to the states . . . or to the people." Unlike federal power, which is granted, the state already has the power, unless expressly or implicitly denied by state or federal constitutions.

Each state is sovereign to itself with its own system of laws and courts, except where state and federal powers interfere or conflict. Where there is conflict, state laws must yield to federal acts.

Although there is a great diversity in court systems among the states and between the states and the federal government, there are basically two types of courts—*trial courts* and *appellate courts.*

A trial court hears and decides controversies by determining facts and applying appropriate rules. This procedure may be carried out before a judge in the absence of a jury, or it may involve a jury. Federal and state constitutions guarantee the right to a trial by jury unless waived by all the parties.

In a trial with a jury, the decision-making functions are divided between the judge and jury. The judge generally decides the questions of law, while the jury generally decides the questions of fact. In product liability cases, mixed law-fact questions arise in which case the jury is called upon to establish the standard for the product, and the function of the court is to provide guidelines for the jury determination of defect. In this way, the legal standard is set for the case under litigation.

At the state level there exist a tremendous diversity of names for trial courts (also called courts of original jurisdiction). There are circuit courts, district courts, county courts, municipal courts; all of these are often lumped together under the general name of *lower courts.* At the federal level, the United States District Courts are the trial courts.

Appellate courts review the decisions of trial courts. If either party (plaintiff or defendant) in a civil suit is dissatisfied with the decision of the trial court, he may appeal to a higher court, the appellate court. The power of these courts of appeal is confined to a review of potential or alleged errors made in the trial court below it. No new arguments or proof are admissable because the appellate court does not retry the case; there are no witnesses or jury. The court bases its decision solely on whether the lower court acted in accordance with the law.

The ultimate court of appeals, or court of last resort, is the United States Supreme Court. It exercises both appellate and original jurisdiction. Within the pluralistic court system of the United States, the Supreme Court is unique in that it reviews certain decisions of lower federal courts.

Only in a few circumstances can a decision be appealed to the Supreme Court as a matter of right. The court controls its docket and reserves its time for important cases that the Supreme Court justices deem to be of far-reaching impact.

#### Jurisdiction

If the engineer finds himself involved, possibly as an expert witness, in a case being tried in a federal district court, he will naturally wonder why the case is in a federal rather than a state court. The answer to this question lies in the area of jurisdiction. Jurisdiction is the power or authority of a court to hear and make judgments in controversies between parties. This power is conferred either by law or constitution.

A complete discussion of jurisdiction would be highly technical, complex, and extensive and is beyond the scope of this book. In this section, jurisdiction will be discussed only in the context of attempting to make the court system more coherent and understandable to the engineer. Furthermore, the discussion here will be confined to a few general concepts. Specific examples of jurisdiction of the various courts will be given in the sections devoted to the court systems.

Jurisdiction can be classified into two general types: jurisdiction over the *subject matter* and jurisdiction over the *person.*

Jurisdiction over the person deals with the problem of bringing the defendant into court and is of little concern to the engineer. In any event, it does not contribute to a clarification of the interplay among the courts or of questions of movement across state or national boundaries.

Jurisdiction over the subject matter, on the other hand, helps to clarify the roles of the various courts. As a general standard, the dictum "to each his own" can be used—that is, federal courts for federal matters and state courts for state matters. Unfortunately, these are not always both necessary and sufficient conditions, so exceptions have to be considered, as will be discussed later.

Jurisdiction over the subject matter of a suit can also depend upon the amount of money involved in the litigation. For example, a municipal court may be limited to claims of less than $10,000, and a justice of the peace (a court not of record) may be limited to some stated claim, such as $300. These dollar amounts vary from state to state. The subject matter of these so-called inferior courts may also be limited. For example, a probate court deals only with wills, the management of estates of decedents, letters of administration, adoption, and mental competency.

In summary, three general rules can be discerned. First, federal matters may generally be brought only to the federal courts; second, local matters will be brought to the appropriate state court; and third, federal courts do not review state court decisions.

### The Federal Court System

A schematic diagram of the federal court system is presented in Figure 2.1. Of the various courts shown, only the U.S. district courts and the U.S. circuit courts of appeal are of interest in connection with product liability.

#### *United States District Courts*

The district courts of the United States are the courts of original jurisdiction (that is, the trial courts) in the federal system. It is in these courts that suits are started, and it is here that issues of fact are determined, either before a judge or before a judge and jury. These federal district courts handle most of the cases that come to the federal courts.

Districts have been established by dividing the nation into 94 geographical territories (districts). Each state constitutes at least one district having a court

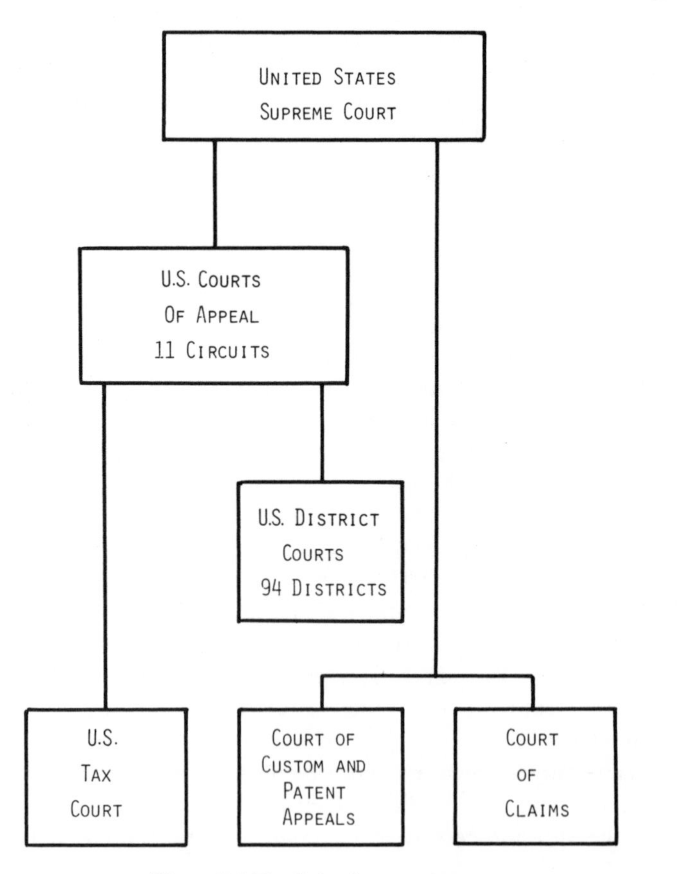


**Figure 2.1 The federal court system**

located in it; some states have two or more district courts. Most of the large cities of the United States have a district court; each court has one or more judges.

In civil actions the federal district courts have jurisdiction in two major categories. The first, and most obvious, category is that involving a strictly federal question—such as might arise under the Constitution or might involve controversy under federal statutes. The second, and more important from the standpoint of product liability, is that involving *diversity of citizenship.*

Diversity of citizenship cases are those in which the amount of the controversy exceeds $10,000 and is between citizens of different states. A citizen is an individual, corporation, or similar entity. A corporation is considered a citizen in the state in which it is incorporated and in the state of its principal place of activity. It cannot sue or be sued in a federal court in either of these states if the other party is also a citizen of either of these same states; in this situation the action would be brought to a state court.

The federal district courts are not permitted to entertain jurisdiction over suits for less than $10,000. If, for example, Jones of New York wishes to sue Smith of Ohio for $5,000, he must bring his action in a state court. If he wishes to sue for more than $10,000, he will have the option of bringing his suit either in a state court or in a U.S. district court.

Quite often the litigants have a choice of bringing their suit in a federal court or in a state court. If the suit is brought in a federal court, the court is bound by the statutes and precedents of the state in which the district court sits. This avoids different results being reached by a federal court and its associated state courts.

Virtually all product liability suits that are tried in federal courts have been brought under the rule of diversity of citizenship.

#### *United States Circuit Courts of Appeal*

The United States is divided into 11 geographical circuits with each circuit containing several states. Sitting in each circuit is a U.S. court of appeals which hears appeals or questions of law from decisions made by the U.S. district courts. In most cases these appeals courts are the ultimate appellate tribunal in the federal system, because appeals to these courts are a matter of right (which is not true of the Supreme Court).

The circuit courts of appeal have no original jurisdiction; they will only consider questions of law or error from the lower trail courts (the district courts).

### The State Court Systems

The court systems of the states are similar in their functions, though they differ widely in the details of their structure. All states have inferior and appellate courts. All have a system of local or county courts of original jurisdiction, and

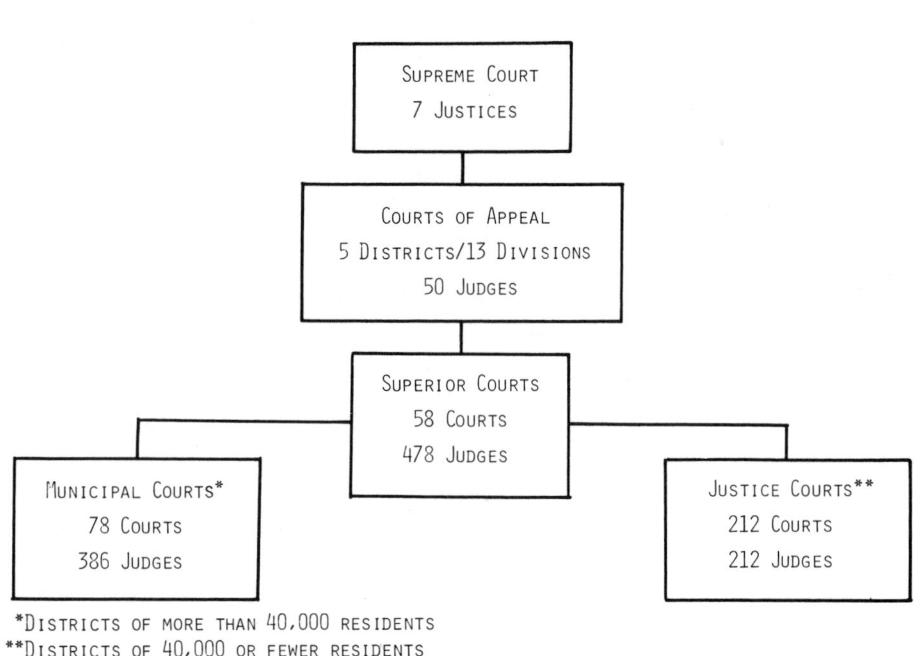


**Figure 2.2 A typical state court system**

all have courts of final appeal. And nearly every state has a minor judiciary consisting of magistrates and justices of the peace.

The dissimilarity among state court systems consists in the great variety of names used to designate courts actually having similar jurisdiction. For example, trial courts of original jurisdiction may bear the name of common pleas, district, superior, circuit, or even supreme court.

A state judicial system usually includes a probate court to handle estates of the deceased, a small claims court for low maximum amounts of claims, local courts (municipal, city, county) of limited jurisdiction, and trial courts. A schematic diagram of a typical state court system is shown in Figure 2.2.

## Legislation

To the average citizen, the word *law* means substantive law. This is the legal process in which statutes are enacted, either defining new law or adopting the common law. This is the primary responsibility of the legislative branch of government.

In other words, the legislative branch creates law by enacting statutes, while the judicial branch creates law through case decisions. To phrase this slightly differently, the legislature declares what the law shall be, while the courts declare what the law is.

The function of the courts in construing acts of the legislature is called *statutory construction.* In exercising this function, the courts do not have legislative authority. That is, the courts must not construe statutes more broadly, or give different meaning, than was intended by the legislature. This is particularly true when the statute is clear and unambiguous. Furthermore, the courts never consider abstract or moot questions of law; they decide only real controversies.

In a world in which legal processes are becoming more and more pervasive, the role of the legislative function in law is becoming more and more important. This is due in part to the unwieldy and haphazard nature of the common law system. In the modern world there is far greater need for certainty and uniformity in laws, both among the states and within federal legislation.

In general, industry and business in the United States do not conform, and are not confined, to the rather artificial boundaries of state lines. Yet for many years there were preposterous conflicts and contradictions among the various state laws dealing with interstate commercial transactions. Although the problem was recognized as early as 1900, it was not until 1945 that the efforts of hundreds of lawyers, judges, and law professors were focused on the task of creating a coherent integrated statute to deal with commercial transactions among the states. The result was the *Uniform Commercial Code* (UCC), which was first adopted by the State of Pennsylvania in 1954. By 1972, this code had been adopted by all the states except Louisiana.

The UCC replaced many isolated statutes which dealt with such things as banking processes, sales, warranties, receipts, documents of title and lading, and stock or other securities transfers.

### Administrative Law

With the tremendous growth and impact of technology since World War II, there has been a corresponding growth in government administrative agencies that regulate and monitor the operations of business and industry. Every citizen and business is affected by these administrative agencies as much as or more than they are by the judicial system.

There are a large number of federal and state administrative agencies. One of the earliest federal agencies was the Interstate Commerce Commission (ICC), created in 1887. Other well-known agencies are the Federal Trade Commission (FTC), the Internal Revenue Service (IRS), the Food and Drug Administration (FDA), and the Environmental Protection Agency (EPA).

More recently, manufacturers have been subject to controls and procedures developed by the agency administering the Occupational Safety and Health Act (OSHA) and by the Consumer Product Safety Commission (CPSC). These agencies have the objective of protecting the safety and health of workers and consumers.

Although these administrative agencies are separate from the executive, legislative, and judicial branches of government, they exercise all three of these functions to some extent. These agencies are empowered by the legislature, within certain defined limits, to exercise powers delegated by the legislature. These powers include the issuing of regulations, investigations, rule making, enforcement, and adjudication. A great deal of expert knowledge is required in this process which is not generally possessed by the courts.

There are many legal problems posed by the existence of both agencies and courts. For example, the power to investigate is one of the functions that distinguishes agencies from courts. This power is usually exercised in order to perform properly another primary function, and like any other power of government, it must be exercised by the agency in accordance with constitutional rights.

Administrative agencies have the power to issue regulations which have the force of law and which cover a very broad range. This power to make, alter, or repeal rules and regulations is called rule-making power; it is legislation on the administrative level which is confined to the granting statute. In exercising this power the Federal Administrative Procedure Act (APA) requires that notice of proposed rule making be published in the *Federal Register* not less than 30 days before the effective date of the rule.

Administrative agencies have adjudicatory power, which involves the determination of rights, duties, and obligations of persons. Adjudicatory hearings are sometimes called quasi-judicial hearings, because they resemble court hearings, However, in an administrative proceeding the decisions are made by administrators rather than judges; they perceive their function as one of implementing a legislative purpose rather than hearing the cases of two litigants. However, these decisions may be appealed to the courts.

The courts recognize that they should interfere as little as possible in the determinations of administrative agencies because of the expertise that the agency represents. This is called judicial self-restraint and comes from recognizing that agencies and courts as coordinate instrumentalities of justice. On the other hand, the judiciary will step in when it feels that an administrative agency has acted arbitrarily in matters of law. It is understood that a reviewing court considers questions of law and not of facts.

To some extent the distinction between law and fact is one of function rather than definition. If a court believes a given question lies within its area of competence it will call the question one of law. If, however the question lies within the competence of the jury or fact finder, the court will call it a question of fact.

Arguments against administrative agencies usually take the position that the administrative process of adjudication is not truly judicial, because the agency is interested primarily in forwarding its own objectives. This disadvantage is balanced by the advantage an agency has in being expert in some particular ares. It should be remembered that problems brought before the regular courts are infinite in variety.

# 3 State and Federal Laws and Regulatory Agencies

## Introduction

For many years, the common law of both England and the United States obligated the employer to minimize hazards and prevent personal injury or property damage to both employees and customers. Nevertheless, it was not until early in this century that an employee could realistically expect to be compensated for an injury resulting from his employment. Beginning around 1900, the working public began to demand that employers recognize the worker's broken body, as well as broken machinery, as a legitimate expense of doing business. This resulted in legislative action, at both the state and federal level, which led ultimately to *Workmen's Compensation* laws.

More recently, consumerism and the protection of the public in all situations have become important political forces. This has had direct impact on the engineer through the activities and rulings of federal regulatory agencies such as the *Occupational Safety and Health Administration* (OSHA) and the *Consumer Products Safety Commission* (CPSC). These agencies came into existence because legislators perceived a neglect of safety for workers and consumers. Their basic mission is to protect the public from unreasonable risk of harm.

This new protective climate is epitomized by the rapidly changing body of law that creates rights and duties of both workers and consumers—product liability. The engineer concerned with product liability will need to understand its relationship with the activities of federal regulatory agencies and with workers' compensation laws.

## Workmen's Compensation

There are two basic environments within which an injury or loss could lead to product liability litigation. One of these is the workplace environment where an employee may suffer injury as a result of work he performs for his employer—for example, as a result of operating a production machine. The other is the general consumer environment, where an individual may suffer injury from using a consumer product—for example, as a result of operating a lawnmower to cut the grass at home.

For injuries occurring to employees in the workplace environment, Workmen's Compensation applies and does essentially two things. First, it provides compensation to the injured person (and his dependents); second, it protects the employer from liability which might otherwise result from the injury. As an example, Article II, Section 35, of the Ohio Constitution states the following:

> For the purpose of providing compensation to workmen and their dependents, for death, injuries or occupational disease, occasioned in the course of such workmen's employment, laws may be passed establishing a state fund to be created by compulsory contribution thereto by employers and administered by the state, determining the terms and conditions upon which payment shall be made therefrom. Such compensation shall be in lieu of all other rights to compensation, or damages, for such death, injuries or occupational disease, and any employer who pays the premium or compensation provided by law, passed in accordance herewith, shall not be liable to respond in damages at common law or by statute for such death, injuries or occupational disease.

The Workmen's Compensation Law of Ohio and its administration is defined in a bulletin prepared by the Bureau of Workmen's Compensation.[1] The Ohio law is compulsory and is administered by the state. This is not true in all states. (Note that Workmen's Compensation is not a federal law.) For example, some states do not require compulsory participation. Under an elective law, the employer has the option of not participating; in this instance the employer will also lose immunity from liability actions. The number of such companies rejecting participation is small, and it is estimated that approximately 80 percent of all workers in the United States are protected by Workmen's Compensation.

Each of the 54 Workmen's Compensation laws is different.* Some states require that the employers insure their risks through funds administered by the state, while others permit insurance coverage through private insurance companies.[2]

Workmen's Compensation is really a form of no-fault financial protection. That is, the employer is required to compensate an injured employee without any question of negligence even being considered. The concept was originally

---

*In addition to the 50 states, there are laws for the District of Columbia, Puerto Rico, federal employees, and maritime workers and longshoremen.

that the worker would be certain to receive some scheduled benefits and paid medical expenses in return for giving up the uncertain possibility of recovering more by suing the employer for negligence. The employer, on the other hand, gave up his uncertain defenses based on contributory negligence, assumption of risk, and negligence of a fellow employee.

Although the injured employee gives up his right to sue his employer, he does not give up his right to sue a third party. The third party could be the designer, manufacturer, or seller of the machine that caused the injury. This is how product liability actions arise from the workplace environment. The incentive to sue the manufacturer of the machine by which the worker was injured is that potentially much higher compensation might be received than can be recovered from Workmen's Compensation. From the recovery made from a third party, a portion is generally used as repayment for the Workmen's Compensation account for medical expenses and as a lump-sum payment for the loss of faculties. This will depend, however, on the jurisdiction, because in some states the employer may have to share in providing compensation to the injured worker if the employer has been negligent. Even if the worker decides not to pursue this course, he may have to transfer (subrogate) his right to sue a third party to the Workmen's Compensation insurer. Thus, many product liability suits evolve from Workmen's Compensation cases.

One of the major factors in the present product liability crisis is that the Workmen's Compensation benefits have deteriorated in the face of spiraling inflation. Also, Workmen's Compensation does not compensate for pain and suffering; it compensates only for what is believed to be the injury's effect on loss in lifetime earning capacity and for medical care.

### Occupational Safety and Health Act and Administration

In December 1970, Public Law 91-596, known as the Occupational Safety and Health Act (OSHA), was signed by the President. This landmark legislation has the fundamental aim to ensure "so far as possible every working man and woman in the nation safe and healthful working conditions and to preserve our human resources." This federal law was passed due to inadequate state laws, and it has teeth. Under the law, employers must provide their employees with a safe working environment and protect them from their own carelessness.

Acts such as OSHA serve as excellent examples of the growth of administrative law resulting from the desire to protect the public. This act covers an estimated 57 million workers and 4 million businesses. OSHA not only defines standards of safety for the workplace but includes a very powerful enforcement program to which an employer must comply or suffer penalties that include substantial fines and even imprisonment. In order to administer the act, several new government agencies and organizations had to be established.

The Occupational Safety and Health Administration (which also has the acronym OSHA) was set up to issue standards and administer the act, which has the force of law. To assist OSHA in developing standards, another organization, the National Institute for Occupational Safety and Health (NIOSH), was established within the Department of Health, Education and Welfare (HEW). The function of NIOSH is to carry out research and education in the area of safety and health and to identify hazardous substances.

Every engineer should understand the intimate relationship between OSHA and standards. Under OSHA, the Department of Labor has the responsibility for issuing and enforcing occupational safety and health standards. Since the act was signed, the Secretary of Labor has promulgated consensus standards previously adopted by the American National Standards Institute (ANSI) and the National Fire Protection Association (NFPA). (The subject of standards will be discussed from the point of view of reducing product liability risk in Chapter 6.)

As every engineer is probably aware, OSHA sets workplace standards and is authorized to act to protect the worker only in his place of employment. It is important to recognize the difference between a workplace standard and a product standard. A workplace standard sets the rules and regulations concerning the total environment of an employee when he is at his job. Some of these workplace standards have nothing to do with products—for example, aisle requirements for in-plant safety.

A product standard, on the other hand, is one designed to cover the requirements for some specific product—for example, punch presses. Product standards may be related to workplace standards, because OSHA states that the purchaser of production equipment cannot permit employees to use noncomplying equipment in the workplace. Thus, although OSHA has no authority to restrain a product manufacturer from making or selling noncomplying equipment, it has control of its final use. This is, in fact, a most effective control. Thus, even though the designer is not legally required to follow OSHA standards in designing products, the marketplace will ultimately obligate him to do so.

The design engineer has an obligation to be knowledgeable about those OSHA regulations that affect the products he designs. This will require a thorough knowledge of existing ANSI and NFPA standards, as well as keeping informed about potential new standards or regulations. Proposed regulations are published by the *Federal Register*, and it becomes required reading for any design engineering organization interested in actions potentially affecting the viability and marketability of specific products.

There are several major areas of OSHA regulations and standards related to the design of machinery. One is in the area of electrical requirements. Since the National Electric Code (NEC) is the primary basis for ANSI standards, which in turn are the primary base for OSHA standards, electrical requirements will generally be met if the design complies with the NEC.

A second major area is controls and point-of-operation safeguards, and a third is machine guarding.[3] Finally, to the extent that a production machine contributes to environmental qualities, such as noise and air quality, the designer must be aware of the pertinent OSHA requirements.

What is the relationship of OSHA to the primary subject of product liability? It should be apparent that the same activities required to handle the interfacing with the federal regulatory agencies are specifically involved in the product liability situation.*

While adherence to a consensus or mandatory standard will not necessarily constitute a strong defense for the product manufacturer hit with a product liability action, it should be obvious that noncompliance with OSHA standards will make it virtually impossible to defend against a product liability claim. It should also be obvious that one of the initiation points for regulation will be product liability litigation pointing to real or possible product hazards.

### Consumer Product Safety Act and Commission

In October 1972, Congress passed Public Law 92-573, known as the Consumer Product Safety Act (CPSA). This legislation was designed to provide protection for the public in the nonworkplace environment—that is, to "protect the public against unreasonable risks of injury associated with consumer products." The primary goal of this act is to substantially reduce injuries associated with consumer products in or around homes, schools, and recreational areas.

In May 1973, the Consumer Product Safety Commission (CPSC) was activated to administer and enforce the act. The commission establishes mandatory safety standards for consumer products sold in the United States, and it has the legislative authority to enforce these standards in the courts with civil penalties of up to $500,000 in fines and criminal penalties of up to $50,000 in fines and one year in jail.

Congress has directed the CPSC to assist consumers in evaluating the comparative safety of consumer products; to develop uniform safety standards; to minimize conflicting state and local regulations; and to promote research and investigation into the causes and prevention of product-related deaths, illnesses, and injuries.

The commission also maintains an injury information clearinghouse and operates the National Electronic Injury Surveillance System (NEISS), which monitors 119 hospital emergency rooms nationwide for injuries associated with consumer products. The objective of NEISS is to identify those products that by

---

*OSHA and CPSC both have the elimination or reduction of *unreasonable* risks of injury as primary objectives. This parallels the primary focus of strict liability, which is that of an *unreasonably dangerous* product.

design or in performance create a hazardous condition in normal use. Although NEISS has found that team sports is by far the most prolific injury producer, this is followed by engineered products such as bicycles and playground equipment. Since 1975, some 28 deaths and 100,000 injuries have been caused by skateboard accidents alone, according to NEISS.

The CPSA stipulates that the safety regulation of certain products is to remain within the jurisdiction of prior agencies—motor vehicles, aircraft, boats, medical devices, foods, drugs, firearms, alcohol, tobacco, and pesticides. The commission is also precluded from regulating risks associated with OSHA.

Since CPSC is the all-encompassing agency dealing with consumer products, it is the one the design engineer must be prepared to deal with on a continuing basis. The engineer should be aware that the commission has the right to review any "new" product prior to its distribution in the marketplace. The "new" designation applies in cases in which there is insufficient information to assess a product's safety or a totally new material or process is used.

The commission also has the power to require a manufacturer to (1) repair products without charge, (2) replace products, and (3) issue a refund.

### Impact of Regulatory Agencies on Design

There are a number of ways in which the design engineering profession can respond to government's concern for safety. First, the designer must keep informed of current requirements by becoming thoroughly familiar with the *Federal Register* and the literature made available by the regulating agencies. Second, the engineer can become actively involved in the standards-setting process by direct participation or through the efforts of professional societies.

In the final analysis, the question is not entirely one of greater competence or skill but rather of an attitude of professional behavior. The engineer must actively assume responsibility and accept accountability for the products he or she designs.

### References

1. *The Workmen's Compensation Law of Ohio, 1973,* published by the Industrial Commission of Ohio. Available from the Bureau of Workmen's Compensation, 65 S. Front St., Columbus, Ohio, 43215.

2. For an excellent survey of the history and diversity of Workmen's Compensation, see W. Hammer, *Occupational Safety Management and Engineering,* Prentice-Hall, Englewood Cliffs, N.J., 1976.

3. R. Sutter, *Mach. Des. Mag. (Jan. 11, 1973);* R. W. Wagner, *ASME Paper 73-DE-10* (1973).

# 4

# Basic Concepts of Product Liability Law

In order to function effectively as a designer or manager, the engineer must understand some of the legal ramifications of his work. This means understanding some fundamental legal terms and doctrines particularly as they relate to safety and product liability.

It is not an objective of this book to develop esoteric theories and definitions of law. This chapter, therefore, is limited to a discussion of the basic concepts of product liability law. These concepts are those of negligence, warranties, and strict liability in tort. These three theories of product liability, in turn, are based upon more primitive legal concepts such as due care, reasonable prudence, foreseeability, and so forth.

The common law bases underlying product liability change as judicial interpretations are modified to meet changing social situations and responsibilities. Nevertheless, certain legal definitions and doctrines continue to hold over a long period of time. The most important ones employed in product liability are defined in Table 4. 1.

## Liabilities—Contract and Tort

Only two fundamental legal bases underly the evolution of product liability. They involve the liabilities of a manufacturer arising from either the principles of contract law or those of tort law.

### *Contract Concepts*

A *contract* is a binding agreement, for whose breach the law provides a remedy. In the context of product liability, a contract will relate to the sale of a product. Such commercial transactions are governed by the *Uniform Commercial Code*

TABLE 4.1

Some Fundamental Legal Terms

---

*Great care:* The high degree of care that a very prudent and cautious person would undertake for the safety of others. Airlines, railroads, and buses typically must exercise great care.

*Reasonable care:* The degree of care exercised by a prudent person in observance of his legal duties toward others.

*Duty of care:* The legal duty of every person to exercise due care for the safety of others and to avoid injury to others whenever possible.

*Liability:* An obligation to rectify or recompense for any injury or damage for which the liable person has been held responsible or for failure of a product to meet a warranty.

*Negligence:* Failure to exercise a reasonable amount of care or to carry out a legal duty which results in injury or property damage to another.

*Contributory negligence:* Negligence of the plaintiff that contributes to his injury and at common law ordinarily bars him from recovery from the defendant although the defendant may have been more negligent than the plaintiff.

*Express warranty:* A statement by a manufacturer or seller, either in writing or orally, that his product is suitable for a specific use and will perform in a specific way.

*Implied warranty:* An automatic warranty, implied by law, that a manufacturer's or dealer's product is suitable for either ordinary or specific purposes and is reasonably safe for use.

*Privity:* A direct contractual relationship between two persons or parties, such as between a seller and a buyer. If A manufactures a product that is sold to dealer B, who sells it to consumer C, privity exists between A and B and between B and C, but not between A and C.

*Tort:* A wrongful act or failure to exercise due care, from which a civil legal action may result.

*Strict liability in tort:* The legal theory that a manufacturer of a product is liable for injuries due to product defects, without the necessity of showing negligence of the manufacturer.

*Proximate cause:* The act that is the natural and reasonably foreseeable cause of the harm or event that occurs and injures the plaintiff.

(Table 4.1 continued)

*Foreseeability:* The legal theory that a person may be held liable for actions that result in injury or damage only where he was able to foresee dangers and risks that could reasonably be anticipated.

*Obvious peril:* The legal theory that a manufacturer is not required to warn prospective users of products whose use involves an obvious peril, especially those that are well known to the general public and that generally cannot be designed out of the product.

*Assumption of risk:* The legal theory that a person who is aware of a danger and its extent and knowingly exposes himself to it assumes all risks and cannot recover damages, even though he is injured through no fault of his own.

*Standard of reasonable prudence:* The legal theory that a person who owes a legal duty must exercise the same care that a reasonably prudent person would observe under similar circumstances.

*Res ipsa loquitur:* The permissible inference that the defendant was negligent in that "the thing speaks for itself" when the circumstances are such that ordinarily the plaintiff could not have been injured had the defendant not been at fault.

*Prima facie:* Such evidence as by itself would establish the claim or defense of the party if the evidence were believed.

*Subrogation:* The right of a party secondarily liable to stand in the place of the creditor after he has made payment to the creditor and to enforce the creditor's right against the party primarily liable in order to obtain indemnity from him.

*Deposition:* The testimony of a witness taken out of court before a person authorized to administer oaths.

*Discovery:* Procedures for ascertaining facts prior to the time of trial in order to eliminate the element of surprise in litigation.

---

(UCC), which has been adopted by every state except Louisiana. The UCC defines "contract" as "the total legal obligation which results from the parties agreement as affected by [the code] and any other applicable rules of law."

The concepts of contract law pertinent to product liability are those of *privity of contract* and *breach of warranty*. Privity of contract means the direct relationship existing between two parties to a contract, such as that between a seller and a buyer. If A manufactures a product that is sold to dealer B, who sells it to consumer C, contract privity exists between A and B and between B and C, but not between A and C.

A warranty may be express or implied. Both have the same legal effect and operate as though the seller had made an express guarantee. An express guarantee is governed by the principles of contract law. Under the UCC, warranty has been contract oriented, as explained above. Nevertheless, warranty in its relationship with product liability has a hybrid origin in both contract law and tort law. This had caused much confusion about the meaning of the expression "strict liability." This will be discussed in more detail later.

An *express warranty* is a part of the basis for a sale; that is, the buyer has purchased the goods (or product) on the reasonable assumption that they were as stated by the seller. No particular form of words is necessary to constitute an express warranty; the seller's statement may be written or printed as well as oral. It is sufficient that the seller assert a fact that becomes a part or term of the transaction between the parties.

Inflated sales talk ("puffing") by the seller cannot be treated as an express warranty when it merely represents a commendation of the product or involves an opinion of the seller which the buyer should not believe. To say that "this product is the best on the market" is merely to give an opinion and would not constitute an express warranty.

Although breach of an express warranty might be the basis for a product liability suit, of far more importance in product liability development is the concept of an *implied warranty*. An implied warranty is one that is not made by the seller but rather is implied by law. This warranty arises automatically from the fact that a sale has been made.

Of the various product attributes guaranteed by an implied warranty, the implied warranty of merchantability or fitness for normal use is of particular importance in product liability. This warranty states, in effect, that the product, by being offered for sale, is reasonably safe for use. This implied warranty for safety is always in effect.

### *Tort Concepts*

A *tort* is a wrongful act or a failure to exercise due care resulting in injury, from which civil legal action may result. A tort is often defined as a civil wrong independent of contract. Tort law establishes standards of human conduct and of duty, for whose breach the law provides a remedy.

The law of torts seeks to provide compensation to members of society who suffer losses because of dangerous or unreasonable conduct of others. The tort theory of *negligence* is one of the most important in the context of product liability.

Negligence occurs when one person fails to fulfill a duty owed to another or acts with less care than would a reasonable and prudent person under the circumstances. Absence of an intent to cause harm is a distinguishing characteristic of negligence. For the tort of negligence to be recognized as a cause of action, two elements must be present—a duty or standard of care recognized by law and a breach of the duty or requisite care—and the breach of duty must be the proximate cause of harm or injury.

The standard for negligence is what a "reasonable" person would have done. The elusive, imaginary reasonable person and the parallel legal concept of unreasonable risk (or danger) are concepts that the courts have grappled with for centuries. A comprehensive discussion of this topic is presented in the Appendix.

### Liabilities—Strict and Absolute

In recent years, "product liability" has become virtually synonymous with *strict liability in tort.* As this suggests, a plaintiff injured by a product and seeking recovery under this doctrine would bring suit in tort. Ordinarily, this would signify a failure to exercise due care or the unreasonable conduct of other persons.

However, the distinguishing attribute of strict liability in tort is the unique feature that the conduct of the manufacturer or the care he exercises is irrelevant. In order to recover damages, the plaintiff does not have to prove that the manufacturer was negligent. He must prove only that the product was defective, unreasonably dangerous, and the proximate cause of harm. In other words, the focus is on fault in the product rather than on fault of the manufacturer. The manufacturer is said to be strictly liable because his liability does not depend on his own conduct or care. This makes defense particularly difficult and frustrating to the manufacturer, and this legal doctrine is the basic cause for what is called the "product liability crisis."

It should be noted that strict liability in tort is not the same as *absolute liability,* because the plaintiff must still prove that the product was defective in the sense of being unreasonably dangerous. The litigation will rest heavily on this pervasive concept of what constitutes unreasonable risk and unreasonable danger. In absolute liability, the existence of injury proximately caused by the product would be sufficient proof for recovery of damages from the product manufacturer. In other words, under absolute liability the manufacturer would become the insurer for any damage arising from his products.

The doctrine of strict liability in tort most commonly relied upon is that based on the *Restatement of Torts,* Second Edition 402A, as follows:

1. One who sells any product in a defective condition unreasonably dangerous to the user or consumer or to his property is subject to liability for physical harm thereby caused to the ultimate user or consumer, or to his property, if
   (a) The seller is engaged in the business of selling such a product, and
   (b) It is expected to and does reach the user or consumer without substantial change in the condition in which it is sold.

2. The rule stated in subsection (1) applies although
   (a) The seller has exercised all possible care in the preparation and sale of the product, and
   (b) The user or consumer has not bought the product from or entered into any contractual relation with the seller.

Strict liability in tort is similar to implied warranty liability in that the defendant is liable from the fact that his defective product caused harm and that the absence of the manufacturer's negligence will be no defense. Strict liability in tort is an attempt to avoid some of the technical arguments that developed during litigation under implied warranty theories. These problems include the concept of privity, whether the contractual statute of limitations or the negligence statute applied to a particular case, and whether an actual purchase was necessary to protect the ultimate consumer.

When implied warranty is stripped of the contract doctrines of privity, disclaimers, and its inconsistencies with express warranty, it is a doctrine almost indistinguishable from strict liability in tort. Yet it must be remembered that implied warranty is an action brought in contract rather than in tort, so that in this sense it represents *strict liability in contract.*

If the expression "strict liability" is used alone, it could be considered an amalgamation of negligence and implied warranty theories, because it would not indicate whether the liability is in tort or contract.

### Causes for Product Liability Actions

As has been discussed above, there are essentially three theories under which liability is imposed on the suppliers of products in the United States—negligence, breach of express or implied warranty, and strict liability in tort. These theories of product liability are not mutually exclusive. Thus, a given set of facts may give rise to two or more theories of liability—for example, breach of warranty and negligence.

In many cases, the circumstances will dictate the theory under which the suit for damages is brought. For example, a warranty liability is barred if injury is sustained after the time period defined by the statute of limitations, which runs from the date of delivery of the product. This would not prevent bringing a suit under the theory of strict liability in tort, because under this theory the statute of limitations runs from the date of injury.

Similarly, if the action is brought under the theory of negligence, contributory negligence on the part of the plaintiff will provide a strong defense to the manufacturer. For this reason, most product liability actions today are brought under the theory of *strict tort* (strict liability in tort), because contributory negligence is generally not a valid defense in most jurisdictions. This situation is in a state of flux, because the courts are increasingly indicating that *comparative fault* will apply in product liability cases. In a jurisdiction that follows the comparative fault rule, the plaintiff's recovery will be reduced in proportion to the fault he contributes in the accident.

Although breach of express or implied warranty has played an important role in the development of product liability theory, today the manufacturer is more likely to be sued in tort. This means either negligence or strict tort, the basic difference being that in negligence the litigation focuses on the conduct of the manufacturer (duty and care), whereas in strict tort the litigation focuses on the properties of the product (defective and unreasonably dangerous).

Under either of these two tort theories, there are similar areas in which engineering and management are vulnerable. These are the following:

1. Design
2. Manufacturing and materials
3. Packaging, installation, and application
4. Warnings and labels

As an example, consider the design area. A plaintiff could claim either that the design was not done with the exercise of due care (negligence) or that the design was defective (strict tort).* In the former case the conduct of the designer would be questioned, whereas in the latter case only the designer's results would be questioned.

### *Design Liabilities*

Liabilities stemming from design, whether they are imposed on the basis of negligence or strict tort, are usually based on the following premises:

1. A concealed danger has been created by the design.
2. Needed safety devices have not been included in the design.
3. The design involved materials of inadequate strength or failed to comply with accepted standards.
4. The designer failed to consider possible unsafe (hazardous or dangerous) conditions due to abuse or misuse of the product which were reasonably foreseeable by him.

---

*Note that in this sense, the product consists not only as hardware but also as a design.

The design engineer is directly concerned with potential liabilities posed by these four premises. In addition, he is indirectly involved with manufacturing, packaging, installation, and developing warnings. For example, the designer may foresee a possible dangerous misuse of the product for which there is no realistic design solution. In this situation, it is incumbent on the designer to work with other departments in the organization to draft suitable warnings and labels to make vividly known that the product is potentially hazardous.

Despite the considerable efforts of a designer, can a product be said to be defective because it could have been designed better? At present the law does not require that a product be "accident proof" or so encumbered with safety devices and other appurtenances as to make it noncompetitive. A manufacturer is not necessarily liable merely because his product could have been constructed in a way that would have made its use less dangerous. Here again, the test arising in litigation will involve the concept of what is reasonable or unreasonable. This means invoking concepts such as the benefit-cost ratio and risk-utility theory, which are discussed in the Appendix.

### *Instructions and Warnings*

The total responsibility accompanying product design, development, and marketing includes more than just the duty to provide a safe, functional design. The manufacturer and the engineer have the additional duty to provide instructions and directions for using the product and to warn of dangers involved in using the product. Failure to do so can result in liability under the doctrines of negligence or strict liability or because the implied warranty of merchantability would not be satisfied in the absence of necessary instructions or warnings. To have any effect, instructions and warnings must warn of the dangers inherent in the failure to follow instructions or warnings.

Foreseeable dangers, uses, and misuses play an important role in product liability litigation, and the supplier of products increasingly has the duty to foresee even the rather extraordinary uses of his product. Furthermore, if use of the product without instructions or warnings would expose the user or bystanders to unreasonable risk, the product is, in effect, defective. For these reasons, manufacturers have resorted increasingly to the "warn against it" approach where potential defect problems exist.

It is naive to believe that warnings are the solution to latent design defects. Where unreasonable danger of the product can be cut drastically by redesign, the courts may decide that the manufacturer will not enjoy immunity from liability simply because he decided to cut the risk by using warnings.[1]

For some product designs, the duty to warn may require warnings so sharply worded and severe as to curtail marketability and utility of the product. In litigating product cases of this kind, a court decision for the injured plaintiff on failure-to-warn grounds constitutes, ipso facto, an instruction to the manufacturer to redesign the product in a manner so as to reduce sharply its danger level. Thus, the relationship of warnings and product design is an intimate one, and they are not easily separated.

For certain kinds of products, various state and federal laws regulate the subject of instructions and warnings. If a defendant has failed to furnish such instructions and warnings, he may be liable for violation of statutes as well as subject to common law product liability for the harm caused.

### The Concept of Defect

Product liability under the doctrine of strict liability in tort is predicated on a resolution of the issue that the product is "in a defective condition unreasonably dangerous to the user or consumer. . . ." Product liability litigation properly revolves around the meaning of the associated questions of unreasonable danger and of defective condition (or defect).

It is a source of uncertainty and confusion for the engineer that it is not possible to give precise definitions of "defective" and "unreasonably dangerous." In fact, only a jury or judge can ultimately decide whether or not a product was defective and unreasonably dangerous in a specific situation.

A discussion of the concept of the "reasonable man" and the parallel concept of "unreasonable risk or danger" is presented in the Appendix. It concludes that the unreasonably dangerous defect can only be given substance in terms of a subjective amalgam of appropriate risk-utility considerations.

The relationship between defect and unreasonable danger has been a troublesome one for the courts. This relationship is quite dependent on the kind of defect that exists in the product. The product may have a design defect, a production defect, a warning or labeling defect, or some other kind of defect.

The risk-utility considerations discussed in the Appendix apply basically only to one kind of design defect, namely, a defect resulting from *conscious* design choices. In such cases, the courts have ruled that "defective" means "unreasonably dangerous" and that the test for the jury will therefore be based on risk-utility balancing using considerations similar to those listed in the Appendix.

It is important to be able to distinguish among the various kinds of defect situations and what they imply. In a true production defect situation, the plaintiff will have to establish only that an identified production flaw precipitated the product failure and that the failure caused injury. Risk-utility theory has no relevance in this situation, because the true production defect case is one in which the product does not meet the manufacturer's own internal production standards. Furthermore, a product's physical defect does not necessarily mean that every product in a product line is defective.

Design defects are of a different nature. Design defects can result in condemning at least one feature of each and every product in an entire product line. Design defects may result either from inadvertent design errors or from conscious design choices. The courts have noted that like production defects, inadvertent design errors are subject to measurement against a built-in objective standard. Conscious design choices, however, necessarily involve trade-offs among safety, utility, and cost. Therefore, the concept of defective condition alone is inappropriate and cannot be discussed meaningfully without relating it to unreasonable danger. A more complete discussion of these concepts is included in the Appendix and in Ref. 1.

## Reference

1. A. S. Weinstein et al., *Products Liability and the Reasonably Safe Product,* John Wiley & Sons, New York, 1978.

# 5

# Product Liability Evolution

The idea of product liability can be traced as far back as 1800 B.C. when Hammurabi, the king of Babylon, compiled his great code of laws. This code provided for stringent measures against a craftsman guilty of making a faulty product. The penalties were so severe that the offense was considered more of a criminal than a civil nature. The injured party received no compensation other than the satisfaction derived from knowing of the severe punishment of the offender.

As new products were developed down through the ages, the general rule of caveat emptor—persisting from Roman law—prevailed. This philosophy of "let the buyer beware" was generally accepted by buyer and seller alike as a fair basis for transaction. Of course, products were simpler and their operation more easily understood in those days. Also, scarcity of products discouraged imposing penalties on manufacturers whose wares, even if not safe, were sorely needed.

Around 1900 several important changes began to take place. The productive capacity of the country exceeded the need for simple, basic products, and manufacturers set up product development departments to create more sophisticated products utilizing the growing potential of electrical power and chemicals. Persons injured by these products began to sue the manufacturers, and the courts began to make exceptions to the rule of caveat emptor. During the next 70 years, product liability law evolved from precedent-setting judgments called landmark decisions. The evolution of product liability law can, in fact, be traced by reviewing these landmark decisions.

### Historical Perspective and Landmark Decisions

The possibility of recovery by a plaintiff due to the negligence of a defendant existed at the very beginning of our industrial age. However, product liability actions based on negligence were narrowly restricted by the ancient doctrine of privity. Under this narrow legal concept, lawsuits were admissible only between parties having a direct contractual relationship (privity of contract)—normally, the buyer and seller. Users other than the purchaser of the product could not bring suit against the seller, and the purchaser could not sue the manufacturer if he was not the seller of the product.

An early decision in England was the basis behind this idea that there could be no liability unless there was privity. This was the case of *Winterbottom* v. *Wright* (1842), in which the English court held that the injured driver of a defective mail coach could not maintain an action against the supplier of the coach, because no privity of contract existed between them. The English court felt at that time that a decision in favor of the plaintiff would "be the means of letting in upon us an infinity of actions."

Exceptions to the privity doctrine began to develop in cases in which the product was found to be inherently dangerous to life or health. In 1916, the death knell of the privity doctrine in negligence cases was sounded by the landmark case of *MacPherson* v. *Buick Motor Co.* (1916). MacPherson was driving a new automobile when one of its wheels fell off, causing injury to the plaintiff. The automobile manufacturer's defense was based on lack of privity and the fact that the defective axle was purchased from another manufacturer. Judge Benjamin Cardozo, who later became a Supreme Court justice, held that the manufacturer's liability did extend to the user, because "If the nature of a thing is such that it is reasonably certain to place life and limb in peril when negligently made, it is a thing of danger; and if to the element of danger there is added knowledge that the thing will be used by persons other than the purchaser, then the manufacturer of the thing of danger is under a duty to make it carefully." This famous decision put aside the notion that the duty to safeguard life and limb when the consequences of negligence may be foreseen grows out of contract and nothing else. One vital requirement of the plaintiff remained, however, namely, that he must prove that the manufacturer was negligent.

Proving negligence is often a difficult thing to do, because the burden of proving that the seller or manufacturer failed to exercise due care is a heavy one. If, for example, the plaintiff was contributorily negligent in causing his own injury, no recovery can usually be obtained under the negligence doctrine in most jurisdictions.

Under the aggressive attack of plaintiff's bar, the requirement of negligence began to erode. At first this involved simply bringing suit under a different doctrine, namely, for breach of implied warranty. A warranty, unlike negligence, is not a tort concept but rather a contract concept. Therefore, negligence of the manufacturer does not have to be proved.

Despite its apparent ease as compared with negligence as an effective basis of recovery for the plaintiff, implied warranty was still encumbered with the contract concept of privity. In 1960 the privity doctrine was again struck down—

this time for implied warranty cases—by the famous landmark case *Henningsen* v. *Bloomfield Motors, Inc.* Shortly after purchasing a new car from Bloomfield Motors, Mrs. Henningsen was driving at a speed of approximately 20 mph when she suddenly heard a loud noise under the hood and felt the steering wheel spin in her hands. The car veered sharply to the right and crashed into a wall. Witnesses corroborated her testimony; however, the front of the car was so badly damaged that it was impossible to determine if negligence on the part of the manufacturer was involved to any degree. Suit was successfully brought under breach of implied warranty of merchantability and fitness, even though there was no privity of contract between Henningsen and the defendants. The Henningsen court stated that "where the commodities sold are such that if defectively manufactured they will be dangerous to life or limb, then society's interests can only be protected by eliminating the requirement of privity between the maker and his dealers and the reasonably expected ultimate consumer."

In addition to striking down the privity doctrine in an implied warranty case, the Henningsen court declared the standard automobile disclaimer, advanced as a defense, to be invalid on the basis of ambiguity and because of the inequitable bargaining position of the consumer versus the automobile industry.

After *Henningsen* v. *Bloomfield,* it became apparent that every product placed on sale would have a warranty, implied if not expressed, guaranteeing that it is safe unless specific warnings are provided to indicate otherwise. Also, any defective product that could be dangerous subjects the manufacturer to liability, even if he has exercised care in producing it or has issued a disclaimer of liability in advance. This case made clear that "the burden of losses consequent upon use of defective articles is borne by those who are in a position to either control the danger or make an equitable distribution of the losses when they do occur." These conditions collectively are strikingly similar to strict liability in tort, and the Henningsen case can be viewed as the precursor to the doctrine that the manufacturer is strictly liable for defects in his products.

A major change in legal doctrine resulted from the continuing effort of plaintiffs' attorneys and the courts to insure that the injured product user could have legal recourse at least equal to the technical defense capabilities of the manufacturer. In 1962 a landmark case recognizing strict liability in tort for the sale of a dangerously defective product was *Greenman* v. *Yuba Power Products, Inc.*

The circumstances that brought about the decision in this case were that the plaintiff, Mr. Greenman, was seriously injured when a piece of wood, which he was turning on a popular combination lathe-saw-drill press machine, flew out of the machine and struck him on the forehead. He brought suit for damages against both the retailer and the manufacturer, alleging negligence and breach of warranty in selling a product improperly designed because of inadequate set screws.

The trial court ruled that there was no evidence of negligence, and it found for the plaintiff based on breach of warranty. Upon appeal, the California Supreme Court criticized the warranty basis of the trial court's decision as being unnecessary. The Supreme Court stated that "a manufacturer is strictly liable in tort when an article he places on the market, knowing that it is to be used without inspection for defects, proves to have a defect that causes injury to a human being." The court also stated that "the purpose of such liability is to insure that the costs of injuries resulting from defective products are borne by the manufacturers that put such products on the market rather than by the injured persons who are powerless to protect themselves. Sales warranties serve this purpose fitfully at best."

The Greenman case was the culmination of two historical trends in product liability cases. One was the increasing tendency of the courts to view their mission as one of providing compensation to injured parties as opposed to assessing fault. The other trend was to provide a tort action for the plaintiff free of defendant defenses based on questions of negligence, privity, disclaimers, warranties, and so forth.

Thrusting the burden on those most able to pay, without regard to fault, hardly seems to make for equity under the law. At best, it seems apparent that the courts, in adopting the rule of strict tort, were attempting to correct what they perceived to be a social problem rather than to remedy some deficiency in the law. In any event, strict tort is the liability theory likely to be used in the great majority of states. The engineer must recognize this fact and act accordingly.

The ultimate in manufacturer's liability would be absolute liability, in which the mere fact of injury would be sufficient proof for damages to be awarded to a plaintiff injured by a product. This relief from identifying the product defect is also part of another doctrine, called *res ipsa loquitur* ("the thing speaks for itself.") The doctrine is designed to permit a jury to infer negligence or defect from the occurrence of the injury in cases in which there is no specific evidence pointing to the defect but negligence or defect is the most reasonable inference to be drawn from the facts of the case. Most courts treat res ipsa loquitur as a form of circumstantial evidence. It is an evidentiary doctrine and not a rule of strict or absolute liability. It permits a court to draw a factual inference when that is the most logical explanation.

A good example of this doctrine is represented by the verdict rendered in the case of *May* v. *Columbian Rope Co.* The plaintiff was using a rope-and-truss arrangement on a scaffold. The rope was one-half inch in diameter, three strand, manila line. It broke and May fell from the scaffold. The plaintiff's case was built on the fact that the rope was new and that the truss did not damage the rope. However, the plaintiff did not name a specific defect as the cause of the accident.

The defense explained the testing procedures used in rope manufacture and questioned the alleged newness of the rope and the way in which it was used. The defense also suggested that Mr. May well might have contributed to his own misfortune.

The court said simply that "the thing speaks for itself", that is, the rope broke, caused an accident, and injured the plaintiff. May recovered damages.

This theory is often applied in medical malpractice suits. For example, if a sponge is left inside a paitent after an operation, the mere fact of its presence indicates negligence.

As threatening as res ipsa loquitur is, it appears not often to be successfully invoked. This is because there are relatively few cases in which one product can unequivocally be assigned full cause of an accident. Typically, two or three adverse factors are involved in product-related injuries. For example, many products are installed by contractors who are independent of the manufacturer. When an accident occurs, the question usually arises of whether it was caused by a defect in the product or faulty installation. This uncertainty is enough to negate the claim that "the thing speaks for itself."

### Foreseeable Use and Misuse

In addition to the famous landmark cases, there are other important cases from which fundamental lessons can be learned. In general, two important principles will be emphasized. The first is that the product must be designed and manufactured with its total life and total use environment in mind, that is, from the moment it leaves the factory until it no longer exists. The second principle is that the manufacturer is responsible not only for accidents caused by the intended use of the product but also for those caused by other foreseeable uses and misuses. "Foreseeable" is a key legal word.

Perhaps the potential accident easiest to overlook is one that takes place before the product is put to use. What can happen while the product is still in its package? Here is an example. A door manufacturer was transporting doors overseas by ship. The doors had a large opening in the top half for glass to be installed later. A number of them were packed together, forming a stack 42 inches high. A cardboard cover was then placed around the center for protection and two steel bands used to secure the stack. The cardboard did not cover the ends, which gave the appearance of solid wood; however, the openings in the doors were in line and this void covered by the cardboard. The cardboard had no marking on it other than the words "fine doors."

A longshoreman carrying a 100-pound sack of flour walked across the doors and sustained injuries as he fell through the void area. In the ensuing trial the manufacturer argued that this was a clear case of abuse. However, the Ninth Circuit Court of Appeals affirmed the judgment that the injury was caused solely by the manufacturer in packaging the product. The manufacturer should have known that it is customary for stevedores to walk on material already loaded. Damages were paid because of failure to foresee that a product can cause injury even in transit.

Another easy-to-overlook aspect of product life is the physical changes that take place in the materials that compose it due to normal use and environment. As a case in point, consider the following.

A farmer suffered loss of one eye when a piece of metal chipped off a hammer he was using. This was a forged-head carpenter's hammer, and it was being used to drive a pin into a clevis to connect a manure spreader to a tractor. Both sides agreed that there were no metallurgical flaws when the product left the manufacturer. However, a process known as work hardening occurs whenever metal is squeezed, struck, or bent. Also, it was agreed that work-hardened metal is more likely to break off in chips when striking an object harder than itself. At the time of the accident the hammer, which had work-hardened, had a Rockwell hardness of C-52 and the clevis pin a hardness of C-57. Thus, by mere use of the hammer and by the coincidence of the clevis pin being harder than the hammer, the necessary conditions for an accident to occur existed.

The defense lawyer argued that the hammer was not being used as intended and that the farmer should have used a ball-peen hammer for this job. However, it was pointed out that the farmer bought the hammer at the local hardware store and that this type of hammer is well known to be used for all types of tasks.

The court found in favor of the farmer on the basis that work hardening is well known to metallurgists, that the manufacturer's records showed chipped hammers had been returned to the factory for replacement, and that the manufacturer had the duty to foresee that the hammer might be used in the way in which the farmer used it.

The important lesson is that all materials change with age and environment. Electrical insulations deteriorate, metals oxidize, lubricants congeal, and so forth. In thinking about what type of accidents can occur during the use or foreseeable misuse of a product, one must include in the scenario, all of the adverse changes that typically affect the materials that make up the product.

It is also necessary to guard against accidents occurring to a person who is not involved in the use of a product. Consider the case of the little girl who attempted to get on the seat behind the operator of a riding mower. The mower blades were well guarded for 270 degrees around the front and sides, but the rear was entirely exposed. The girls's foot was badly mangled. This case was settled out of court for a considerable amount of money before the case came to trial.

In considering all foreseeable circumstances, operator neglect must be given due concern. For example, a self-propelled lawn mower was left unattended. The design of the control lever was such that it shifted from "disengage" to "engage" because of the vibration of the running motor. The mower started to move, struck the plaintiff, and caused serious damage to his right foot.

Another case in which an unattended product caused damage involved a vaporizer that overheated after the water had boiled away, causing a fire. The defense claimed this was misuse of the product in that the plaintiff failed to follow the manufacturer's instructions by allowing all of the water to boil away. However, the court held that it was too much to expect a consumer to use a vaporizer properly all of the time. Occasional failure to refill it when it became dry was inevitable. Furthermore, since an inexpensive automatic cutoff switch would have obviated the need for perfect user compliance, failure to have included this protection constituted negligence in design.

So far we have emphasized the unusual situation. There are also many cases in which accidents happen during normal use. For example, in another vaporizer case the unit was designed with a loose-fitting cap to avoid high pressure being developed by the steam. A three-year-old child suffered third-degree burns after knocking such a vaporizer over. The defect in design was alleged to be the loose-fitting cap. It was pointed out that the cap could have been attached with molded threads and small holes used to allow the escape of steam. This would have prevented the massive amount of hot water from escaping as it did.

In another case a motorist was returning home after shopping when suddenly the engine of his car stopped. While checking the battery he struck a match for light, and the gas from the battery exploded. The plaintiff charged defective design and an inadequate warning label. (The latter will be discussed later). There was a substantial award, but it is not known which of the two-part allegation carried more weight with the court.

A case that points up the need for product designers to be familiar with the variations from the norm in people's size and shape was decided by a federal District Court in Florida. An award of over $5 million was made to a high school student injured during a football game. The plaintiff alleged that the rear edge of the helmet pressed against his neck as he was tackled, causing permanent paralysis. Considering the differently shaped heads and necks of football players, it is probably impossible to design a helmet that is proper for all players. This case dramatized the dangers of improperly fitting football equipment. As a result, most high school athletic personnel have taken much-needed steps to ensure proper choice of equipment for each player.

## Summary

Articles describing liability cases with startling verdicts appear in newspapers, popular magazines, and technical journals and give sufficient details to allow you to monitor the current situation and future trends. As we have said, the law is an ever-changing discipline.

We have described how product liability arrived at its present state of strict liability and have reported sufficient cases to emphasize the need of the designer to study the entire life cycle of the product, from the moment it leaves the factory until it no longer exists. During this total life the engineer must consider every use and foreseeable abuse to which the product might be subjected. He must consider the results of chance failure of all parts and of failure due to wear. He must consider injury to the user and injury to anyone who might happen to be near. He must consider injury to a person of typical size, skill, and intelligence and to the atypical person.

Modern product design demands a broad view of the responsibility of the designer. The designer must not only be talented but must also be responsive to human welfare. Litigation should always be a last resort; good product design means minimizing the risk assumed by the product user.

# 6

# Reducing Product Liability Risk

With the advent of the theory of strict liability as a basis for product liability actions, it may appear that little can be done to prevent legal action against a manufacturer's products. Nonetheless, there are many things that the engineer and management can do to lessen the likelihood of product liability suits.

The main countermeasures are adherence to principles that will ensure design and production of truly safe products. Use of these policies will protect both the manufacturer and the public. They involve both a broadening of the design perspective and a more professional approach to the design process.

Traditionally, design engineering has revolved around strictly technical considerations, such as strength of materials, machine elements, electrical considerations, manufacturability, and costs. In an era in which the manufacturer is extremely vulnerable to product liability action, this traditional approach must be broadened to include more attention to standards and codes, hazards analysis, express and implied warranties, warnings and labels, failure analysis, and design review.

For some companies this will require substantial changes from design procedures currently used. There is no doubt that many products of the past were designed by persons who had little or no formal engineering training. These products were designed simply by specifying what appeared to be a product having reasonable strength, reasonable electrical insulation, reasonable heat-transfer properties, and so forth. Many such designs were verified by running tests on one model made by a craftsman, which was not representative of those made for sale by mass production.

In undertaking activities to lessen the risk of product liability, the first necessary action for success is for the top management of the company to formulate company policy statements and to identify a person of sufficient authority to be responsible for product safety. Then a formal listing of product liability management activities should be prepared. Some of these activities may place constraints on the product that will be unpalatable to marketing, purchasing, or other departments. They may also require additional expenditure of time and effort by engineers, technicians, and nontechnical personnel. Thus, without sufficient backing and commitment by management, the activities are not likely to be carried out.

### Twenty Guidelines for Product Liability Management

Product liability management involves a formal overview of an entire program of activities designed to lessen product liability risk. Some of these activities are included in the engineer's formal training. Others, however, may be new to the practicing engineer. An example is the preparation of warnings and labels.

With the proper endorsement by management, the climate is created for everyone to become conscious of the need to give product safety its proper priority. In this climate, the engineer responsible for developing a product should carry out the following activities.

1. Include safety as a primary specification in identifying the needs during all phases of the product's existence. Books on product design describe the life of a product as consisting of production, distribution, consumption, and discard or salvage stages. Engineers are accustomed to determining in detail what product is needed to meet competition and what characteristics (functional, physical, aesthetic) the product must have. This is a very early step in any design procedure. These specifications are principally set by the third stage, namely, consumption, or use of the product. The product must be safe not only during that stage but during other stages also. Even in the discard stage the product must not poison streams, must not explode, and so forth.
2. Design to a nationally recognized standard. A more complete explanation of this will be given later.
3. Select materials and components that are known to have sufficient quality and a small enough standard deviation from the norm to do the job expected consistently. Material characteristics change with the environment. Thus, of two materials being considered, the one with the greater yield strength at 25 degrees Celsius may have the lesser yield strength at operating temperature. Or, of two components for sensing displacement, one may have a significantly higher reliability than the other. Information about material and component characteristics can be found in volumes of reference books for virtually every alloy and every plastic. These plus

manufacturers' publications on components should be available either as part of a company library or as part of your personal collection.

4. Apply accepted analysis techniques to determine if all electrical, mechanical, and thermal stress levels are well within published limits. The present-day engineering education is largely made up of courses in analyzing various physical situations. This advice merely calls for the design engineer to use those skills that he acquired in such courses to the fullest extent. This makes mastery of all courses in the engineering curricula not only relevant but necessary.
5. Test the device using accelerated aging tests. If possible, a recognized test should be used. If there is no recognized test available, the engineer should develop an accelerated test that truly represents in-use conditions at elevated stress levels. Designing such a test is indeed difficult and usually evokes dissenting opinions as to its value. However, the test is certainly valuable if it helps the engineer discover a defect in the design. Even in the absence of such positive results, the test is valuable in disputing testimony of design defects if the product should become involved in a liability suit.
6. Make a failure and hazards analysis of the product for each stage of product life given in number 1. This will be discussed in more detail later.
7. Make a worst-case analysis of the product by assuming that the material characteristics, part dimensions, and so forth will take on values at the tolerance extremes that are detrimental to the product performance. It is highly unlikely that all parts will simultaneously assume the worst values. However, if this condition is approached, the engineer should be satisfied that the product is not hazardous.
8. Make sufficient information available to the factory (notes on drawings, component specification, etc.) to eliminate hazards (for example, no burrs on this side, hardness must not exceed B85 Rockwell, plating must not chip at this point, etc.). As simple as this sounds, it is necessary. A note about the need to avoid burrs on one side may cause a die to be made one way rather than another. Rolled copper sheets or strips have grain and above a certain hardness may crack when bent along the grain but not if bent perpendicular to the grain. Slight burrs, a slightly damaged formed part, and many other seemingly unimportant things that distinguish mass-produced parts from craftsmen-made parts can result in unsafe products.
9. Make a permanent record of the history of the product development, giving sufficient information on all activities from the needs analysis to

the pilot production run. This should be so complete that others can give reasons for each design decision after the engineer is gone or if he is busy with other responsibilities.

10. Conduct a design review that includes persons knowledgeable about distribution, installation and use, manufacturing problems that might arise to lower the quality of the product, and dangers to persons and to the ecology after discard. In the preceding chapter we discussed the case of the door manufacturer who did not know that stevedores walk on materials in the holds of ships. Other cases of damage occur from rough handling during loading and unloading trucks and from the vibration the product is subjected to during transportation from normal road irregularities. For example, the bracket that holds a small transformer in a control unit may break by fatigue from such vibration, even though it is certainly strong enough to hold the transformer during normal use.
11. Inform the person in charge of quality control of possible manufacturing errors that can result in a dangerous product. In a sense, this is merely an extension of number 8. However, it deserves to be emphasized. Tool engineering and production is separate from quality control in most manufacturing organizations, but dies and molds wear, and jigs and gages become misshaped. The quality-control arm of the company must be relied upon to reject parts and assemblies when minor variations will interfere with the product's safe operation.
12. Work with the advertising department to guard against overstatement of product performance. This will be discussed further in the section on warranties.
13. Use warning labels on the product when appropriate. This will be discussed later.
14. Have all products inspected after manufacture (100 percent inspection) where feasible. Functional testing should be used as a final check when possible. It is important that the design engineer be at least a participant in deciding the verification tests to which each device produced for sale should be subjected. It is also important for all to realize that tests at elevated stress levels are at least to some extent destructive. Thus, for example, high-voltage lineman gloves when tested at a higher voltage than that for which they are designed are damaged by that test.
15. Supply unambiguous instructions for properly installing or using the product. This is best done by thoughfully writing the instructions and then having someone no more skilled than the installer interpret them. If difficulty in interpreting instructions occurs in that test, it is likely to occur in the field.

16. Determine any necessary service or maintenance necessary to keep the product safe and operating according to specifications. If appropriate, the engineer should provide a maintenance record form to ensure compliance by the user with the suggested maintenance program. For example, every manufacturer of electrical ground fault interrupters for personal protection insists that the device be checked monthly by pushing a test button. A form is provided with each unit for the home owner to record when the device was last checked.

17. Encourage sales and service personnel and dealers to report any complaints that have to do with injury or economic loss. The report should include the following:

    a. whether a defect in the product is alleged to exist

    b. a clear description of the alleged defect

    c. how the defect is said to have caused the accident.

    d. the nature and extent of the injury or economic loss

18. Submit the product to an independent testing laboratory for their evaluation and approval (or listing) if possible. Some laboratories engaged in this work are described in the section on standards.

19. Test the effects of mass manufacture on the product by having a pilot run made by production personnel using production tools. Random samples from this run should be subjected to the same accelerated life tests and safety analysis as were the experimental samples during the product development. The results of these tests should then be compared to the results of tests made previously on engineering samples. If performance is deficient for the production devices, remedial action should be taken.

20. Document risk-utility considerations made in the design wherever there is a question regarding safety in connection with the product. In other words, if some improvement in safety would be possible only at the expense of the product's utility, the engineer should document the study made. He should also document possible safety improvements that would most probably result in the product's becoming noncompetitive or unmarketable.

The engineer is, of course, only one of the people responsible for product safety. This list reflects what we believe should be done by the engineer. It does not address the relevant activities of the many other employees of the manufacturing organization which are just as necessary for production of safe products.

The following sections discuss in greater detail some of the more important items listed above.

## Standards and Codes

Standards of performance have existed for several thousand years. It is reported that Hammurabi, the king of Babylon in the eighteenth century B.C., stated that if a house collapsed and killed its occupant, the builder would be put to death. Undoubtedly, this regulated building construction effectively. It can be considered a model code from the point of view of emphasizing performance rather than rhetoric.

The difference in the definitions of "codes" and "standards" is slight. In general, the former is used where the system of principles or rules that make up the document are specifically required by law, while the latter is used where at least some degree of consent is involved. In extreme cases, the consent may be academic. For example, many political subdivisions require that equipment used for electrical power distributions be listed by a testing laboratory. Underwriters Laboratories (UL) is the universally recognized testing laboratory in this field. Thus, manufacturers of such equipment follow Underwriters Laboratories' standards because not to do so would severely limit the market for their products.

As another indication of the fine distinction between codes and standards, it is worth noting that each revision of the *National Electric Code* (NEC) is submitted to the American National Standards Institute (ANSI) to be accepted as an ANSI standard. For convenience, we will use "standard" to refer to all documents that are established by legal authority, recognized expertise, custom, or general consent and that state requirements or give advice about the characteristics, the design, or the installation of materials or products.

### *Types of Standards*

There are many kinds of standards. There are workplace standards, product standards, mandatory standards, and voluntary standards.

A workplace standard sets the rules and regulations concerning the total environment of an employee when he is actively working at his job, while a product standard is one designed to encompass the requirements for a specific product. The standards developed by the Occupational Safety and Health Administration (OSHA) are workplace standards, and a product manufacturer cannot obtain compliance of product design through OSHA. In some cases, products can be submitted to independent testing laboratories to determine compliance with product standards.

A mandatory standard is one having the force of law. For example, the various standards adopted by OSHA and published in the *Federal Register* are mandatory standards. A voluntary standard, on the other hand, is usually one developed and promulgated by a trade organization or similar group. Compliance with a voluntary standard is not a requirement of law, even though noncompliance can often result in serious legal problems.

We are concerned here almost exclusively with product standards. Not all such standards are necessarily the same in either content or authority. Thus, although this section deals specifically with safety standards, it should be noted that other types of standards also exist, dealing with such things as interchangeability of parts.

The product design engineer must be sensitive to suggestions from a wide spectrum of different groups. Although an engineer is likely to believe that standards written by professional societies or trade associations have more status than suggestions for product improvements made by consumer groups, a jury involved with a liability lawsuit may be just as willing to accept the one as the other. In fact, if a consumer group criticizes a certain characteristic of a product in their publications, a jury is likely to place full responsibility for any accident resulting from that characteristic squarely on the manufacturer.

Of course, according to the principle of strict liability, the fact that the designer followed a safety standard in developing the product that caused an accident does not relieve the manufacturer of legal responsibility. However, various groups have established safety standards through careful consideration of those same problems that a designer working to reduce product liability must face in developing a product. The fact that following such standards will in general result in a safer, better-quality product is the best reason for using them. Also, since in the majority of product liability cases the causative factors and the chronology of events are seldom known precisely, a successful defense is often based upon convincing the jury that of several possible causes, it is doubtful that the accused product is at fault. Showing that the product conforms to a safety standard is an important aid in establishing such a defense. In fact, the more universally the standard is accepted as representing good design, the more likely the jury will be to exonerate the product.

### *The Content of Safety Standards and Codes*

The requirements imposed upon a product by a standard or a code can be written in only two ways—as prescriptive requirements or as performance requirements. A prescriptive requirement deals with materials and dimensions. For example, according to the Underwriters Laboratories Standard for Cabinets and Boxes (UL 50), steel enclosures for electrical equipment to be used outdoors must be protected by a zinc coating defined as G90 by the American Society for Testing and Materials (ASTM) or by a G60 coating and certain types of enamel. As another example, Underwriters Laboratories Standard for Enclosed Switches (UL 98) requires that the distance between uninsulated parts of opposite polarity be ¾ inch through air and 1¼ inches over surface for voltages between 126 and 250 volts. These are called prescriptive requirements because, in effect, they determine the design decisions rather than permitting the engineer to decide what is adequate.

A performance requirement, on the other hand, deals with a performance test that the product must successfully complete to comply with the standard. These typically involve increasing the electrical, mechanical, or thermal stress on the product and then operating it repeatedly. The combination of increased stress and number of operations is designed to represent worst-case use or to accelerate wearout. For example, representative circuit breakers that are rated at 15 amperes and are of the type used in homes to protect the wires running to the convenience outlets are required by Underwriters Laboratories to be able to interrupt rated current 6000 times, interrupt six times rated current 50 times, then interrupt a current that would reach 10,000 amperes if the breaker were not there three times. It must still be operative after these and the nine or ten other tests that make up the evaluation program.

Performance requirements allow the designer full latitude in finding ways to comply, and from the point of view that they promote creative design and continual product improvement, they are preferred over prescriptive requirements. However, most standards contain both prescriptive and performance requirements.

### *Standards Organizations*

As an engineer concerned with product liability, you will need to locate standards that might apply to your product. We will endeavor to give you sufficient information in this section to start your search. However, it is not possible to give a complete and definitive list. You should be wary of any publication that implies that it does this. The number of available standards in any product area is large and continues to grow. Often the origins are organizations that previously have not been engaged in standards work. Their publications might come as surprises, but once available, they cannot be ignored. For example, in 1976, the Aluminum Association published their *Guide to Specification of Electrical Installations Employing Aluminum Conductors* "to enhance safety and reliability of electrical installations. . . ." This organization had previously published data on characteristics of aluminum alloys; in this area, their expertise is unchallenged. The use of aluminum in electrical equipment, however, is another matter. Nonetheless, the existence of this publication presents the problem that in any legal action involving overheating of electrical conductors, its suggestions may be introduced to support allegations of defective design.

Also in 1976, the Nebraska Inter-Industry Electrical Council, Sprinkler Irrigation Association, and Agricultural Research Service of the U.S. Department of Agriculture approved their *Standards for Electrical Service and Equipment for Irrigation,* whose purpose is to prevent shock hazards. Neither of these organizations is listed in available references of standards-writing agencies.

On the other hand, the problem of acquiring necessary standards is not overwhelming. There are relatively few organizations that account for the majority of standards. Careful surveillance of these organizations, participation in professional society activities, and affiliation with trade associations appropriate for the class of products of interest to you will probably put you in touch with the standards you need.

Another important source of standards information is the *Product Standards Index, Second Edition,* by V. L. Roberts, published by Pergamon Press. This 500-page book identifies many standards-writing organizations and lists standards that apply to various products. For military standards, an *Index of Federal Specifications and Standards* is available from the Superintendent of Documents, U.S. Government Printing Office, Washington, D.C. 20402.

The following organizations can be considered as primary sources of standards.

1. The American National Standards Institute (ANSI) does not write standards. Its purpose is to be the clearinghouse for standards that are written with full participation of all concerned. Many of the standards of other U.S. organizations that we will mention, as well as foreign standards, have been designated as ANSI standards. Thus, for the engineer concerned with product liability, this organization is the first place to look for standards that might apply to a given product.
2. The Underwriters Laboratories (UL) is probably the most widely known standards-writing and testing organization in the United States. Their standards apply to materials and devices and are meant to prevent loss of life and property from fire, crime, and casualty. There are over 350 UL standards, three major testing laboratories (New York, Chicago, and Santa Clara), and offices of local inspectors throughout the United States to provide constant surveillance of products before they leave the factories where they are produced.

   Products that comply with UL standards bear a label or the initials "UL" circumscribed by a cricle as evidence of compliance.
3. The American Society for Testing and Materials (ASTM) develops standards on the characteristics and performance of materials. It publishes over 4000 individual standards in over 30 volumes, and each standard is also available as a separate publication.

   A typical ASTM standard identifies a certain property of a material that is of interest, describes the equipment necessary to carry out the test that quantifies the property, and carefully describes the test procedure.
4. The National Fire Protection Association (NFPA) promotes and improves methods of fire prevention and protection. The NFPA publication list is extensive and touches any subject that conceivably could be associated with fire. The NEC, one of its publications, is accepted as the standard for installation of electrical equipment by most political subdivisions

throughout the country and as such is the source of many product changes that are eventually made in UL standards. NEC is reissued every third year, with changes agreed upon by the members of 21 panels in response to suggestions submitted to them.

Since many accidents involve fires, the NFPA's authoritative *Fire Protection Handbook,* now in its fourteenth edition, and their *Fire Protection Guide on Hazardous Materials* are of special interest to engineers concerned with product liability.

5. The National Safety Council (NSC) devotes its entire effort to the prevention of accidents. Part of its effort is directed toward general educational activities, as indicated by the recent involvement with the National Electrical Manufacturers Association (NEMA) in developing slide tape presentations on the use of ground fault interrupters to prevent electric shock. However, most of its publications are related to the design of chemical plants.

There are certain nonmilitary government agencies that write standards or solicit and support standards written by organizations that may be interested in doing so. The activities of these organizations should be watched carefully, because these government agencies are given the power by Congress to impose their standards with legal action. Two such agencies, OSHA and the Consumer Product Safety Commission (CSPC), were discussed in Chapter 3. Additional such agencies are listed below, but this list is not all-inclusive.

6. The Food and Drug Administration (FDA) is the oldest consumer safety and protection agency. Its origin was the Food and Drug Act of 1906, but today its activities also include cosmetic products. This agency endeavors to use voluntary compliance, but it can rely on court orders to seize a dangerous product. The engineer's involvement with this agency is usually related to food and drug processing equipment or containers that malfunction. You probably remember instances in which the whole nation was alerted to the discovery of botulism in several cans of food which was caused by improper preparation or canning. In these situations complete recall of the offending product is immediately required.

7. The Federal Trade Commission (FTC) is related to product liability mainly through the Magnuson-Moss Warranty Act, which will be discussed later. It also warns the public when it believes a product to be unsafe or dangerous. This has happened in recent years when Christmas tree lights and dolls having eyes of poisonous seeds were imported.

8. The Occupational Safety and Health Act of 1971, which established OSHA, adopts consensus standards such as the NEC or sets its own standards and determines if workplaces comply with these standards. Its inspectors levy penalties if violations are found. Repeated violations can re-

sult in heavy fines and jail sentences. Obviously, any citation issued by OSHA that involves your product would be damaging in a product liability suit. Thus, anyone involved in the design of products that are likely to be used in industrial or commerical installations should know what standards OSHA relies upon in that product area to be certain of full compliance.

9. The Consumer Product Safety Commission (CPSC) was created in 1972 to protect the public from unreasonable risk of injury from consumer products. It is primarily concerned with products used in the home. The first activity of the agency was to determine which product categories most frequently cause injury. They found that "certain makes and models of 16 categories of products subject the consuming public to unreasonable hazard." These categories are, in alphabetical order, architectural glass used in sliding doors, color TV sets, fireworks, floor furnaces, glass bottles, high-rise handlebar bicycles with elongated seats, hot-water vaporizers, household chemicals, infant furniture, ladders, power tools, protective headgear, rotary lawnmowers, toys, unvented gas heaters and wringer-washers.

   One of the problems the CPSC has is to determine just what constitutes a consumer product. The key to distinguishing a consumer product is its ability to provide its function when standing alone. A fire alarm functions independently of being mounted in any particular place and is considered a consumer product even when built into a home by the builder. On the other hand, roofing shingles serve no purpose until properly attached to the roof by the builder and are thus not a consumer product.

   The need to establish jurisdiction has been a source of disagreement in the main task tackled by the CPSC, namely, the problem of overheating of aluminum wire termination in homes, which may cost up to $400 million to correct. Are the products used in wiring a home considered consumer products or not?

10. The Environmental Protection Agency (EPA) was created in 1970 to coordinate federal environmental activities. Authority is given to the agency by the Clean Air Act, Water Pollution Control Act, Safe Drinking Water Act, Solid Waste Disposal Act, Federal Insecticide, Fungicide and Rodentcide Act, Toxic Substances Control Act, and Noise Control Act. The EPA sets environmental quality standards, monitors pollution levels, and sponsors research related to environmental pollution. Its requirements on automobile exhaust pollution are well known to the general public and have been a major influence in automobile design. Its involvement with most companies, however, is to reduce the objectionable effluents from the manufacturing facility. Thus, chemical engineers are most likely to be the ones involved with this agency.

The following two important consumer research groups should be known to engineers involved with consumer products.

11. Consumers Research, Washington, N.J. 07882 and Consumer Reports, Orangeburg, N.Y. 10962 publish reports periodically that discuss the quality of products they purchase from retail outlets, that is, not specially prepared samples. The reports typically include a general discussion of what they believe to be important for customer safety and satisfaction as well as ratings on how the specific devices they tested performed. As was stated earlier, if such a consumer organization criticizes a certain feature of a product and that feature is alleged to have caused injury, a jury is likely to be influenced by their comments just as much as by comments of the professional or trade society standards.

*Summary*

Standards have long been important to the engineer and are even more important now because of product liability. We advise you to write to those organizations that are likely to have standards in your product area. Ask for a catalog of their publications. They usually provide this at no charge, since selling standards is an important source of their income. The study of these catalogs should give you the titles for an adequate start for your library of standards.

Also ask for any publication offered on the history and operation of the organization. Learn how each organization arrives at its standards, what enforcement authority it has, and how compliance with its standard is verified. This information is most important if the opposing attorney tests your knowledge of the organizations when you are acting as an expert witness.

## Hazards and Failure Analysis

A *hazard* can be defined as a condition having the potential to cause harm or injury to people, animals, or property. A *risk* is the probability of an accident occurring when a hazard is present, and *danger* is a combination of a hazard and a risk of serious consequences.

Products can possess both inherent hazards and the potential for contributing to or initiating other hazards. The existence of hazards is determined from experience, analysis, and careful study.

Perhaps the most common type of hazard is that which is an inherent property or characteristic of a product. In an electrical device, for example, electrical shock is often an inherent hazard. In a chain and sprocket transmission, the "pinch point" represents an inherent hazard. Many of these inherent hazards are known from long experience, and a listing of major ones is presented in Table 6.1.

The interaction of people with products introduces hazards associated with human performance and behavior. These can be difficult to anticipate completely, and a partial listing of some of these human hazards is presented in Table 6.2.

The failure of a product or product component can constitute a hazard if safeguards have not been provided in anticipation of failure. Such product failures consist not only of those due to physical causes, such as stress or fatigue,

TABLE 6.1

Inherent Hazard Sources

| *Chemical* | *Mechanical* |
|---|---|
| Corrosive | Weight |
| Toxicity | Speed or acceleration |
| Flammability | Stability |
| Pyrophoricity | Vibration |
| Explosive | Rotation |
| Oxidizing | Translation |
| Photoreactive | Reciprocation |
| Hydroreactive | Pinch or nip points |
| Carcinogenic | Punching, shearing |
| Shock sensitive | Sharp edges |
| | Cam action |
| *Electrical* | Stored energy |
| Shock | Entrapment |
| Short circuit | Impact |
| Sparking | Cutting actions |
| Arcing | |
| Explosion | *Miscellaneous* |
| Radiation | Noise |
| Fire | Light intensity |
| Insulation failure | Stroboscopic effect |
| Overheating | Temperature effect |
| | Pressure, suction |
| *Radiation* | Emissions |
| Alpha, gamma, beta | Ventilation |
| X rays | Ignition sources |
| Infrared, ultraviolet | Decomposition |
| Radio and microwaves | Slipperyness |
| | Moisture |
| | Aging |

TABLE 6.2

Human Hazards

| *Personal* | *Human errors* | *Environmental* |
|---|---|---|
| Ignorance | Failure to perform | Weather |
| Boredom, loafing | Incorrect performance | Noise |
| Negligence | Incorrect supervision | Temperature |
| Carelessness | Incorrect training | Light |
| Horseplay | Overqualification | Floor texture |
| Smoking | Poor judgment | Ventilation |
| Alcohol or drug use | | Complexity |
| Sickness | | Comfort conditions |
| Exhaustion | | Warnings |
| Disorientation | | Social factors |
| Stress | | Psychological factors |
| Physical limitations | | |
| Cultural background | | |

but also those arising because of inadequate consideration of human factors when the product was designed. The evaluation of hazards from product failures needs to be carried out by the designer in an organized and methodical fashion. One such method is called failure modes and effects analysis (FMEA).

#### *Failure Modes and Effects Analysis*

FMEA is concerned with identifying the ways in which a product or product component can fail and the effects of such failures. Each component of the product or system is examined thoroughly, and results are tabulated in a systematic manner. An examination of the failure modes, their symptoms and effects, their probability of occurring, and their injury potential leads to an assessment of a hazard index for each component, assembly, subsystem, or system.

The format of a FMEA varies somewhat depending on the objectives, but generally the following sequence of steps is involved:

1. Description of the component, assembly, subsystem, or system whose failure mode is being identified.
2. Identification and description of the ways in which the component can realistically fail (the failure modes).
3. Determination of the symptoms of each failure mode.

4. Determination of the effect or result of each failure mode.
5. Determination of the probability of occurrence for each failure mode. Statistical data is used where feasible; otherwise, a qualitative ranking can be used, such as 1 = very low, 2 = low, 3 = medium, and 4 = high.
6. Assessment of injury potential associated with each failure mode. Again, a qualitative ranking can be used, such as 1 = no injury, 2 = minor injury, 3 = major injury, and 4 = death.
7. Determination of a hazard index based on the combination of factors in numbers 1 through 6. The hazard index must also include additional considerations, such as the time available to take corrective action and the environment of failure.

A complete FMEA can be organized and carried out in a methodical way by developing worksheets in matrix form in which the tasks itemized above form the columns and the component failure modes form the rows.

The FMEA may range in sophistication from a qualitative engineering judgment to a quantitative finite element analysis. Its value lies in its systematic approach in analyzing product performance in critical detail. It also serves as an excellent basis for conducting design reviews.

The FMEA does not focus on causes or interrelationships existing between the elements of complex products or systems. It answers the question of what will happen if a postulated failure occurs rather than identifying the source of the failure. For complex systems a method called fault tree analysis (FTA) is used to determine how a failure may occur.

*Fault Tree Analysis*

The conventional way of applying a fault analysis to a system of components would be to estimate the probability of each event that can lead to a major failure (or catastrophe) and determine from this the probability of the major failure itself. When carried out in this way, the analysis is extremely time-consuming and unnecessarily complex.

An alternative method is to start with the undesired major failure (perhaps identified from a FMEA). This is called the "top" event. Then the logical combinations of faults occurring at the next level of components or subsystems that could cause the "top" event are determined. This process is repeated at the next lower level of components or subsystems, and it continues down the heirarchy until the most primitive faults have been identified that could lead to the "top" event by a whole sequence of successive faults.

The backward or "top-down," approach is called a fault tree analysis because when it is completed the diagram looks like an upside-down tree with the major failure (or catastrophe) as the base of the trunk, the first level of components as the main branches, the next level as smaller branches emanating from the main branches, and so forth. Although one usually thinks of the life of a tree as flowing from the roots to the tips of the branches, the logic of the fault tree flows in the opposite direction. If failure of the smallest branches can be avoided, the major failure located at the trunk will never occur. This is the value of the diagram.

The construction of a fault tree is actually made using "AND" and "OR" logic gates and the methodology of Boolean logic analysis. When this analysis is coupled to reliability data for the components, it becomes possible to determine (1) what combination or combinations of faults could cause the "top" event and (2) how often the "top" event might occur. Since a product might fail in any number of ways identified from a FMEA, each failure mode ("top" event) requires a separate fault tree. In a complex design the fault tree developed for any "top" event may become so large and complex that the analysis is done using a computer.

As an example, consider a switchboard that distributes electrical power to a factory. The "top" events could include electrocution of personnel, destruction of the building by fire, destruction of the switchboard by short circuits, and so forth. To consider the last, note that a switchboard is made up of an enclosure, bus bar runs, control devices, switches, indicating devices, and probably a transformer. If destruction of the switchboard by massive short circuits is to be studied as the "top," event then a particular situation, say, a short circuit to ground, would be listed as the trunk of the fault tree, and each of the subsystems named above would be drawn as main branches from the trunk. Now considering only a main branch representing the enclosure, a subbranch from it would be the bus bar support structure, another would be the access doors, and still another would be the compartment barriers. Each of these would be examined in detail, that is, to such an extent that we consider whether the hinges on the access doors can fail, allowing the door to fall against the bus bars when the door is opened for inspection or repair. If so, perhaps piano-type hinges should be used to give continuous support to the edge of the door, or perhaps the doors, or bus bars, should be relocated.

In practice, the analysis begun above must continue until all branches have been added to the fault tree and until all the "top" events have been considered. In this way, appropriate action and design changes can be made to avoid failures.

### *Design Safeguards*

Hazards can sometimes be avoided or mitigated by good design or other procedures. A few of the major safeguards—fail-safe design, back-up redundancy, and lockouts, lockins, and interlocks—will be discussed here. Ref. 1 is an excellent source for other measures that can be taken.

Equipment failures are responsible for many accidents. It is natural, then, to study these failures, which will inevitably occur, with the viewpoint of arranging the failure in such a way as to prevent a dangerous situation from occurring. A fail-safe design is one that ensures that the failure will leave the equipment or product in a state in which no injury can result.

Fail-safe designs can be categorized into the following three types:

1. Fail-passive arrangements, in which failure deactivates the system. Circuit breakers and fuses are good examples.
2. Fail-active arrangements, in which failure activates a warning system while safe operation is temporarily maintained.
3. Fail-operational arrangements, in which failure allows the system to continue safely until corrective action is possible.

Back-up redundancy is really a form of fail-safe design in that when a component failure occurs, a second component takes over and substitutes temporarily for the failed component. However, provisions must also be made for the potential failure of the back-up component as well. An example of a redundant system would be mechanical brake linkages superimposed on a hydraulic braking system.

Lockouts, lockins, and interlocks operate based on two basic principles: (1) isolating a hazard and (2) preventing uncompatible events from occurring in the wrong sequence or at the wrong time. A good example of these devices is the lock on an automobile ignition system and steering column.

### *Summary*

In analyzing hazards associated with any product it is beneficial to look at the problem from as many different aspects as possible. The hazards should be considered during special situations, such as maintenance, transportation, or installation, as well as during normal operation. Another way to look at hazards is in terms of the potential severity of injury or loss of property.

All accident situations cannot be avoided, but steps should always be taken to minimize the consequences of failure through warnings, adequate safety factors and safety margins, monitoring, and increased reliability of components. The efforts of the engineer to eliminate or reduce hazards will help not only to avoid costly lawsuits but also to reduce the human misery resulting from injury or death.

## Technical Literature and Warranties

One of the ways in which a person who has sustained injury or financial loss can hope to receive compensation is on the basis of breach of warranty. A warranty is the representation of the character or quality of a product. In order to pursue successfully this legal strategy, it is necessary to prove that the warranty exists, that it was breached, and that the breach was the proximate cause of the injury or loss. The misrepresentation must be known to the purchaser (or lessee or renter) and must have influenced the purchase. This restriction probably has very little effect on the outcome of any case, however. With the passage of time and prolonged discovery, the purchaser may be hard pressed to remember just when he heard or read the laudatory comment about the product that he believes to be the basis for breach of warranty.

Perhaps the most important thing to remember about warranties is that they exist because of every bit of information given to the purchaser by the manufacturer, the seller, and their representatives. The courts do not consider merely the formal written warranty statement. In fact, the very act of offering the product for sale is to represent it as having at least the qualities of similar products, and thus it involves warranty.

### *Express Warranty*

It is convenient to divide warranties into two major classes, depending upon how the product representation is communicated–express or implied.

Express warranties are defined by the Uniform Commercial Code to be the following:

1. Any affirmation of fact or promise made by the seller to the buyer that relates to the goods and influenced the sale.
2. Any description of the goods that influenced the sale.
3. Any sample or model that influenced the sale.

Catalogue descriptions, pictures, advertisements, sales brochures–all affirm facts, make promises of quality, or describe advantages and thus become part of the warranty. Furthermore, there is no limitation that says the affirmation or descriptions must be in print. A statement by an engineer or salesman about a product can also be held to be an express warranty.

Such statements, however, do require individual evaluation as to their seriousness. The courts have recognized that statements are sometimes made during the interchange before a sale that are laudatory but do not constitute a warranty. For example, in selling an electric razor a salesman might very honestly say, "I've used one for years and have had no problem with it." Or a car salesman might say, "I believe this make of car is the best on the market." The first is a statement of fact about one device out of the entire product population. It is not reasonable to conclude that all others wll perform exactly like that one, namely, with "no problem." The second is simply an opinion and must be accepted (or rejected) by the buyer as such. These presale statements are called "sales talk" or "puffing" by the courts. On the other hand, if a buyer is told by the manufacturer's engineer that the contacts of a control switch will not weld, that statement is part of the express warranty.

### *Implied Warranty*

Implied warranties are divided into two subclasses—merchantability and fitness. To explain the first, the courts hold that the mere act of offering a product for sale communicates in a subtle way, and thus implies, that the product is safe and has at least the minimum qualities one would expect of such a product. Most cases relating to breach of implied warranty of merchantability involve allegations that the specific device that caused the loss or injury was not fit for the ordinary use of that type of product. This can happen simply by an unfortunate combination of parts resulting in a "worst-case" product. The plaintiff does not try to prove that the entire population of products is not worthy of being sold but only that a particular one was unworthy.

The second type of implied warranty is fitness, and it generally results when a specific recommendation is made by a representative of the manufacturer or seller in response to a situation described by the purchaser. For example, a customer asked a paint store clerk what to use to remove paint from a floor. The clerk recommended a lacquer thinner. An explosion occurred during its use, and a death resulted. The store paid damages for breach of warranty of fitness.

Like so many product liability situations, there are interpretations that must be considered. In the present instance the Uniform Commercial Code states that the buyer need not bring home to the seller actual knowledge of the particular purpose for which the goods are intended or of his reliance on the seller's skill and judgment if the circumstances are such that the seller has reason to realize the purpose intended or that the reliance exists. In other words, the seller's recommendation of the product need not be precisely verbalized in terms of the buyer's particular purpose in order for the implied warranty of fitness to apply.

### *Preventing Breach of Warranty*

As always, product liability is best reduced by good design. On the other hand, the warranty of any product can be escalated by overstatement to the point that even a well-designed product cannot meet the promises. Everyone connected with the promotion and sale of products must accept the fact that warranty will be breached if the product—every single one of them—cannot do what has been promised either expressly or by implication. The remedy is to have engineers review all promotional material and set guidelines for statements by salesmen. The engineer's job should be to determine that all statements are technically correct and that there are no overstatements of fact. A recent advertisement in a trade journal gives an example of what should be avoided. It showed a picture of a simple tester to be plugged into a convenience outlet to determine the presence of voltage, whether or not an adequate ground connection exists, and so forth. The copy that accompanied the photograph read "Model XXX ground tester is a simple, *foolproof, totally reliable,* device that can check the ground on 3-wire outlets, 2 or 3-wire equipment and tools, 2/3-wire adapters and even 2-wire outlets. Use requires *nothing more* than *plugging* the device into an outlet, using the ground probe and *noticing* which of the three indicator lamps become lighted. A total of ten tests can be performed with 29 different indications for precise troubleshooting." The italicized words need to be reconsidered. "Foolproof" means that the indications cannot be misinterpreted, even by a fool. Certainly they can. Furthermore, nothing is totally reliable, and with 29 different indications from three test lights, the use does require more than "plugging" in and "noticing."

### *Disclaimers*

There is some relief from possible breach of warranty by using disclaimers. However, the courts tend to subject disclaimers to very strict interpretation in cases of economic loss and hold them to be of no value if personal injury from a product defect is involved. Disclaimers must be written with words chosen so carefully that only lawyers skilled in that area of law should compose them.

### *The Magnuson-Moss Warranty Act*

In 1975, Congress passed the Magnuson-Moss Warranty Act in an attempt to clarify for the consumer the extent of any express warranty given by a manufacturer. Basically the act requires that a manufacturer conspicuously indicate that a written warranty for any product costing more than ten dollars is a limited warranty (as opposed to a full warranty) if any of the following are true:

1. A remedy of the customer's problem causes the customer to bear any expense
2. The life of the warranty is limited
3. The compensation to the customer is limited
4. The customer cannot choose among replacement, repair, or refund if a defect exists
5. The customer must follow a fixed procedure to receive remedy for a defect
6. The warranty is implied
7. An unreasonable depreciation is used in fixing the value of a product that becomes defective after some fraction of its expected life

8. Every customer claiming breach of warranty is not given prompt repair, replacement, or refund with cost to him

The act states that nothing in it shall be deemed to authorize the Federal Trade Commission to require a written warranty or to require that they be of a certain duration. This may indicate the strategy that the law will force more manufacturers to follow in the future.

### *Summary*

The most important things to remember about warranties are that they exist as expressed or implied for all products and that the best defense against breach of warranty is to be certain that all communication to prospective customers is devoid of overstatement. Such phrases as "*assure* positive contact," "provide *perfect* protection," "*no* danger of overheating," "*precise* control," and many others that you can add to the list for the type of product you are involved with should be avoided.

As an illustration, the National Electrical Manufacturers Association (NEMA) for many years used names for electrical enclosures based upon their passing certain prescribed tests. Some of the names used were *rainproof, dustproof,* and *explosion proof.* The list goes on, but this is enough to illustrate the point. Water could get into the "rainproof" enclosure, although not above uninsulated live parts; a finer dust than used in the test could get into the "dustproof" enclosure; and "explosion proof" meant that an internal explosion would be contained by the enclosure, but explosions could still occur. These single words are being replaced by code letters, with the explanation that the code means the enclosure has passed the prescribed test. This is certainly in keeping with what has been said above. "Rainproof" can mean various things to the consumer, but the designation 3R (which will probably replace it) is solely and completely defined by whatever test NEMA uses.

## Instructions, Warnings, and Labels

Design responsibility does not end with the factors of cost, functional utility, aesthetics, and marketability. Safety in use of the product requires a total concept of design, including the important factors of advertising, instructions, warnings, and labels. In particular, warnings have an intimate relationship with product design in a legal sense.

The manufacturer or seller of a product assumes the duty to warn of any dangerous propensities of the product. This duty includes the foreseeability of injury measured by the dangerous potentialities of a product and the reasonably foreseeable and intended uses to which it might be put. The duty to warn requires that it be done with a degree of clarity, intensity, and intelligibility sufficient to cause a reasonable person to exercise caution commensurate with the potential danger.

Although warnings are used to protect the manufacturer from liability as well as the user from injury, it must be realized that warnings are not to be viewed as a simple and inexpensive mode of dealing with hazards and risks where redesign of the product can be accomplished without substantially altering cost-utility factors. Many court decisions for plaintiffs on failure-to-warn grounds are tantamount to directing the defendant to redesign the product in order to avoid liability. Thus warnings are related to design, and vice versa.

If a product has obvious dangers, there is no duty to warn, since the danger is a matter of common knowledge. For example, the fact that a sharp knife is capable of cutting a careless user is a matter of common knowledge and is obvious. Similar remarks can be made for guns, blowtorches, and so forth. The hazards that must be warned against are the less obvious ones–inherent, latent, or concealed dangers the manufacturer has knowledge of but the user cannot foresee. Examples are a stepladder whose rear supports will buckle if a very heavy person stands on the very top, a microwave oven that will subject the user to radiation is the door if left open, and a hardened concrete nail that may chip or break when hit with a glancing blow.

Although the duty to warn of obvious danger is not required as a matter of law, it should be remembered that the law changes rapidly, and the interpretation of this duty has been shifting in favor of the plaintiff. It may also vary from one jurisdiction to another.

### *Warning Communications and Labels*

If it is not feasible to design or guard against product hazards that make the product dangerous, then warnings must be included in labels, instruction manuals, and other literature associated with the product. The warnings must be communicated in a clear, complete, unambiguous, and conspicuous manner. Labels must be carefully designed, and instruction manuals for installation, operation, and maintenance should be carefully drafted to stress safe practices and methods. Remember that a label can be the basis for litigation as well as the product itself.

Warning communications should be drafted in the context of the following general considerations.

1. Intelligibility. It must be kept in mind that the user may not read English or his level of education may be lower then average. The manufacturer of a parathion spray was held liable for the deaths of two Puerto Rican field workers of limited education and reading ability, even though the label

attached to the spray had been submitted to and approved by the U.S. Department of Agriculture and went into detail about how the product could be safely used. However, the court held that a warning on lethal chemicals that does not include a skull and crossbones or comparable symbol is inadequate. On the other hand, dramatic symbols should not be overused because they may provide the only way to alert persons who cannot read that a danger exists and overuse may cause the consumer to become jaded to the message.

2. Adequacy. Warning labels should not only warn of the danger but should also tell what can happen and if preventive measures are possible they should be stated. The words must be chosen with great care. For example, a label reading "Do not heat or use near fire" was held to be inadequate when a bottle of nail polish exploded due to the user's lighted cigarette. People do not always consider a lighted cigarette as fire. The warning label should also state the severity of the harm if the warning is violated. It is not enough to warn a user not to drink the contents of a bottle; one must also state what will happen if he does. The warning must possess the degree of intensity required.
3. Completeness. Warning labels should be complete. If a warning label appears on a product, it implies that every unobvious danger is explained. A label that says that drinking the contents of a bottle may cause death is incomplete if contact on the skin will cause dermatitis and this is not mentioned. A label that warns that certain parts of a product must be oiled to prevent malfunction from rust misleads the user if there are other parts whose oiling is also necessary. As a final example, a lawn mower manufacturer was held to be negligent because he did not warn that a self-propelled mower could slip into gear if left unattended with the engine running. Vibration could change the position of the clutch mechanism, but that would not be obvious to the user.
4. Placement and durability. Warning labels are obviously effective only when they get their message to everyone who may get injured by the product. This is a responsibility that lasts as long as the product. Thus, it is not enough simply to have the label available when the buyer examines the product before purchasing it. The label should be durable and permanently attached to the device itself in a position where it is easily seen during normal operation and where it will not be rendered unreadable by normal wear or obliterated by repainting.

An example of a good warning label is found on a new carpenter's hammer purchased recently. The label design has obviously been influenced by the case discussed in Chapter 5 in which a farmer lost an eye due to a chip from a hammer. The label states "Caution: a hammer can be made to chip if struck against another hammer face or other hardened surface resulting not only in damage to the hammer but possibly in bodily injury; safety glasses should be worn."

### *Nonuser Liability*

Warning labels must be used with the honest attempt to prevent accidents. Their message should be given to all who might be injured by the product. This is a difficult task, because injury can occur to infants, to innocent bystanders, and to property of absent owners. An actual case history indicates the degree of difficulty of this task.

An apartment owner brought suit against the manufacturer and retailer of hair rollers that had been used by one of his tenants. The instruction urged the purchaser to use plenty of water when heating the rollers and not to let it boil away. The warning stated that "rollers may be inflammable only if left over the flame in a pan without water. Otherwise they are perfectly safe." The tenant fell asleep after putting the rollers on the stove in a pan of water to heat. The water boiled away, a fire ensued, and extensive damage was done to the building. The court held that the manufacturer should have anticipated that a user might fall asleep, and thus the caution on the roller box was inadequate. With hindsight we can say that the label should have warned against heating the rollers when there was any chance of dozing or becoming busy with other matters and forgetting about them. Also, the possibility of providing a warning card to be given by the purchaser to the owner of the building if the user rented her home may have heightened the awareness of the danger involved. Finally, the closing statement of the warning label ("Otherwise they are perfectly safe.") has the dampening effect of any further consideration of possible dangers. Perhaps most importantly, the case shows that warning a third party about possible danger is so difficult that a product change to eliminate the danger is a better approach. Of course, this might increase the cost of the product, but safer products often cost more.

### *Instruction Manuals*

A product unaccompanied by the necessary instructions on how it should be used will not normally satisfy the warranty of fitness for normal use embraced within the warranty of merchantability. Furthermore, use of the product without instructions may expose the user or bystanders to unreasonable risk of harm. Therefore, it may be essential to provide instructions or instruction manuals with the product.

More than any other single communication, the instruction manual has the potential for describing the product completely. While such manuals are often prepared by technical writers, the ultimate responsibility for their content rests with the product engineer. This is because only the designer knows about the strengths, weaknesses, and hazards the product possesses.

There are usually two types of manuals, an *owner's manual* and a *service manual*. (Remember that both the user and the potential repairman must be protected from harm.) The owner's manual must contain adequate precautionary statements against foreseeable misuse or mishandling that could result in injury to the user. The service manual should indentify potential hazards to the repairman, who might come into contact with high voltage, high pressure, radiation, toxic fumes or materials, and so forth. The repairman must be provided with adequate instructions about how to deal with any of the hazards that may exist.

In designing instruction manuals, all of the factors discussed above must be considered in addition to accuracy, validity, complexity, clarity, and illustrations. Since these manuals serve as documented, prima facie evidence of warnings and instructions concerning the avoidance of hazards, they should be reviewed by the manufacturer's legal staff.

### *Summary*

A product liability action based on failure-to-warn grounds is essentially a negligence action, because it involves the alleged failure of the duty of the manufacturer to warn of either reasonably foreseeable or intended uses of his product. The challenge to the engineer is to determine studiously all the possible ways the user or an innocent bystander could be injured or property could be damaged and then to warn clearly of those product hazards that creative design cannot eliminate.

No matter how well they are prepared, instructions, warnings, and labels cannot in themselves guarantee consumer safety or freedom from product liability suits. The best solution to reduce product liability is still good design.

## Design Review

Design review is a formal, systematic, documented review and evaluation of a product design by a team of specialists not directly associated with the development of the product. The purpose of design review is to assure that the product is safe and reliable, that costs and materials have been optimized, and that the design complies with its specifications and requirements.

### *Participants*

The members of the design review team should have a broad understanding of design and manufacturing as well as of packaging, transportation, installation, use and maintenance, and final disposal. The participants should be knowledgeable in their own area of expertise, and collectively they should be able to investigate performance, product costs, safety, reliability, producibility, environmental effect, maintainability, serviceability, life-cycle costs, human factors, customer needs and reasonable expectations, pertinent legislation, pending litigation, and so forth.

The functional areas that should be represented by a design review team are shown in Table 6.3. From personnel in these areas, the engineering manager responsible for the product usually chooses a chairman who is not associated with the design or the designer. The chairman is responsible for coordinating meetings, meeting schedules, the agenda, minutes, and the final report. The secretary assigned by the chairman should take notes on items discussed during design reviews and record the action items assigned, persons responsible, and required completion dates. All documentation should be approved by both the chairman of the review team and the cognizant design engineer for the product.

The number of participants in a design review session should be kept to the minimum consistent with the skills and experiences needed to review adequately the product in question. In any case, no design review should involve more than a dozen people. These participants must be able to assess independently the design on its merits, respect other members, offer constructive criticism, and accept suggestions.

### *Design Review Meetings*

The chairman sets the time and place for meetings, contacts each participant, and forwards agenda materials to the members at least ten days in advance of the meeting. The first meeting is basically an orientation session at which specific objectives are set and possibly assignments made. One major corporation uses design review checklists from which assignments can be made to the various review participants.[2]

At the review meeting, the participants present answers to the checklist questions along with their recommendations. Using this technique, each participant knows he or she is responsible for specific areas concerning the design; this assures that the review will proceed systematically. Where problems with the design are encountered, the review team may suggest possible approaches for solutions. However, it is not the function of the team to redesign the product!

One of the most important areas of design review is that of high risk, where design alternatives, trade-off studies, critical parts, and reliability considerations should be documented. This can be a most difficult and complex task, but it is of utmost importance in product liability risk prevention. Other important areas are cost, standards, use environment, schedule requirements and delivery dates, test and inspection, and foreseeable misuses.

The tone of the design review meetings must be constructive. Derogatory remarks or a flat refusal to consider changes should not be permitted. It must be remembered that it is the product design and not the designer that is being reviewed, and care must be taken to not impugn the designer's work. The review

TABLE 6.3

Areas of Functional Expertise in Design Review

| Function of group member | Responsibilities | Type of design review PDR | IDR | FDR |
|---|---|---|---|---|
| Chairman | Calls and conducts meetings of group and issues interim and final reports | X | X | X |
| Design engineers (not associated with unit under review) | Constructively reviews adequacy of design to meet all requirements of customer | X | X | X |
| Design engineer(s) (of product) | Prepares and presents design and substantiates decisions with data from tests or calculations | X | X | X |
| Reliability | Evaluates design for optimum reliability consistent with goals | X | X | X |
| Quality control | Ensures that the functions of inspection, control and testing can be efficiently and economically carried out | | X | X |
| Manufacturing engineer | Ensures that the design is producible at minimum cost and schedule | | X | X |
| Field engineer | Ensures that installation, maintenance, and operator considerations were included in the design | | X | X |
| Procurement representative | Assures that acceptable parts and materials are available to meet cost and delivery schedules | | X | |
| Tooling | Evaluates design in terms of tooling costs required to satisfy tolerance and functional requirements | | X | |

(Table 6.3 continued)

| Function of group member | Responsibilities | Type of design review PDR | IDR | FDR |
|---|---|---|---|---|
| Packaging and shipping | Assures that product is capable of being handled without damage | | X | X |
| Marketing | Assures that requirements of customers are realistic and fully understood by all parties | X | | |
| Consultants, specialists on components, values, human factors, etc. (as required) | Evaluates design for compliance with goals of performance, cost, and schedule | X | X | X |
| Product safety, legal, specialists | Assures that design is able to meet known regulations and foreseeable applications safely and complies with legislation, litigation, and patents | X | X | X |

group is advisory only, and all comments and findings should be resolved to the mutual satisfaction of the chairman and the cognizant engineer. The design engineer evaluates the potential of the design review committee's findings with the result that frequently the original design plus suggestions results in a better, safer product.

A report to the engineering manager should include each suggestion and subsequent decision made by the designer. This report should be sent to members of the design review team as well as to other interested parties.

### *Design Review Stages*

Depending on the scope and complexity of the product, one or more design reviews may be made. Generally there are three formal design reviews—preliminary design review (PDR), intermediate design review (IDR), and final design review (FDR).

A *preliminary design review* is held before the initial design is formulated, that is, during the conception and planning stage. The main purpose of this review is to establish communications among the various departments, such as marketing, engineering, purchasing, manufacturing, and management.

The primary purpose of the *intermediate design review* is to check the design against performance requirements and to prepare for long lead-time items. It is carried out after the basic design has been completed but before detailed drawings have been started.

The *final design review* is conducted when materials lists and production drawings are complete and ready for release to manufacturing. During this review, particular emphasis should be placed on producibility, tolerances, instruction manuals, and safety.

The various functional areas involved in the three design stages are indicated in Table 6.3. All information related to the design should be available as back-up for each review, such as design spec's, minutes of design review meetings, trade-off studies, failure mode and effects analysis, cost data, and so forth. A final report should be issued containing the entire review history; this could be vital to a defense of the product in a liability action.

### *Benefits of the Design Review Process*

The potential benefits of the design review process include the following:

1. A strong, documented basis for defending against strict tort, negligence, or warranty actions
2. A better product design from the standpoint of safety, cost, reliability, materials use, maintenance, and manufacturability
3. Fewer post-design changes and fixes and possibly earlier delivery

The total value of the design review process is difficult to ascertain, but there is not doubt that the systematic, methodical, and formal nature of design reviews provides for an openness, breadth of thinking, creativity, and safety that insures greater protection and satisfaction for both consumer and manufacturer.

## References


1. W. Hammer, *Occupational Safety Management and Engineering,* Prentice-Hall, Englewood Cliffs, N.J., 1976.
2. *Design Review Guidelines,* Booklet MB-3284-B; available from Director, Product Improvement, Headquarters Engineering and Development, Westinghouse R & D Center, Pittsburgh, Penn. 15235.

# 7

# Product Liability Insurance

In the previous chapter, measures were outlined for design, manufacturing, and marketing that are believed to result in safer products. Nevertheless, injuries will still occur and claims arise requiring a defense by the manufacturer or seller. An integral part of management's preparedness for such claims is to obtain adequate insurance coverage. The cost of such insurance coverage has been rising rapidly and has precipitated what has been called the "product liability crisis."

The increased cost of insurance coverage for product liability has certainly had a great impact upon virtually every manufacturing company. For many, that cost has become exhorbitant, and for some, product liability coverage has become impossible to obtain.

Since the product liability prevention programs of some of these same small and medium-sized companies often consist of little more than insurance coverage, the resultant prospect of "going bare" (having no insurance coverage) or of liquidating assets to protect investment is a traumatic one.

This explosive atmosphere has created controversy between manufacturers and insurers. Manufacturers, on the one hand, believe that the insurance industry is blundering by doing an inadequate job of defending product liability cases and that it is all too willing to settle out of court. Some manufacturers contend that many of these cases could be won, and the practice of settling does nothing to offset the trend of seemingly always making an award to the plaintiff. Furthermore, this is harmful to the reputation of the manufacturer and contributes to the insurance costs that seem inevitably to increase at every renewal.

Insurers, on the other hand, cite their losses in the medical malpractice field as a reason for spreading increased costs over all manufacturers without regard for the manufacturer's individual experience in product liability litigation. This same practice has been followed by insurance companies with medical malpractice cases, and for this reason product liability is sometimes called "manufacturer's malpractice." The insurance industry also stresses the necessity of manufacturers to do everything possible to assure that products are safe and, indeed, even conducts regular engineering safety inspections of the manufacturer.

Three salient points emerge from this situation. First, some companies still blithely pass off the liability problem as an insurance problem. Second, insurers are too anxious to settle cases out of court. And third, both manufacturers and insurers seem always to be in unanimous agreement that lawyers are the cause of all these problems.

### Escalating Costs of Product Liability Insurance

In March 1977 the National Small Business Association (NSB) issued a statement before the Senate Select Committee on Small Business holding hearings on S. 527. Among other things, the NSB sought to determine the dimensions of product liability insurance problems by selectively sampling the small business community. Typical replies were as follows:

> In the last year our product liability insurance premiums have gone from $18,000 per year to $125,000 per year with only one carrier indicating any interest in the business at all. The recommendation we receive from our insurance broker is that it is very likely that within the next six months this type of insurance may be completely unavailable to machinery manufacturers. At the currently quoted premium level, this represents a charge equal to about 2½ percent of our annual machinery sales.

> Compared to last year, our product liability insurance premium has just been increased 1100 percent. Yet we had no claims or suits against our company–none ever in the entire 25-year history of our company.

> A small machine tool manufacturing client of ours in Connecticut has been faced with a product liability insurance cost increase from $25,000 in 1975 to $150,000 in 1976. The consequences of this are staggering. I realize it is not the insurance company's desire to have these increases but just their prognosis of what it looks like product liability will cost.

It is difficult to pinpoint the actual escalation in premium costs for product liability insurance. For individual manufacturers, premium increases of 360 percent, 440 percent, 625 percent, and 1100 percent are easily cited, and they range all the way up to the astronomical increase of 10,000 percent. One rule of thumb that seems to be fairly common for the maximum acceptable level of premium cost is 1 percent of sales. When the insurance costs rises to 2 percent of sales, it is certainly considered excessive.

The causes of escalating insurance premium costs are obvious. First of all, there has simply been an explosion in the number and severity of court claims, in judgements, and in subsequent defense costs. Even when cases are settled out of court, defense costs are much higher than they were several years ago. The cost of legal defense for product liability claims has increased to the extent that it now accounts for more than 30 cents of each dollar of insurance premium. A second pervasive source of increased costs is the impossibility of applying actuarial techniques to the product liability problem. The exposure to product liability is almost impossible to predict, and the rules seem to be changing every day. This requires underwriters to use instinct rather than statistics.

In addition to the premium cost, insurance coverage imposes obligations on the manufacturer which must be considered in relation to its entire products-loss prevention and insurance programs. These obligations involve the concepts of *occurrence* (and reoccurrence), *product recall,* and *coverage limits.* These are discussed in Ref. 1.

### The Manufacturer

Some of the manufacturer's problems in working with insurance carriers have already been discussed. The major complaints concern the shrinking, or even lack of, availability of insurance and escalating premium costs. In addition, there is the problem of obtaining "gap" insurance. This situation arises when umbrella coverage requires, say, $1 million primary coverage, but the manufacturer is only able to obtain $500,000 in primary insurance. This leaves a gap of $500,000 between the primary and umbrella coverages.* "Gap" insurance is needed to fill this void, and it is very expensive if, indeed, it can be found.

Faced with the prospect of not being able to purchase product liability insurance from established insurance carriers at any price, many small and medium-sized companies are confronted with the choice of either liquidating assets or going uninsured. Exercising the option of liquidating assets (going out of business) may seem to represent a drastic and shocking solution to the problem of not being able to obtain insurance. However, any company that is financially responsible must recognize the potentially disastrous consequences of becoming an uninsured defendant in a product liability action against it. Even if the company successfully defends itself, the costs could be prohibitive; and if unsuccessful, it is certainly conceivable that a small company with, say, $2 million in assets could be totally wiped out. The management must therefore look to the protection of these assets. It should also be remembered that jobs, payroll, and federal and state taxes would be wiped out at the same time. If a small, privately-owned company is experiencing labor problems at the same time, the inability to obtain

*Typically a manufacturer has insurance to a certain limit—say, $100,000—for each plaintiff. Then an umbrella insurance policy—sometimes by a different insurer—covers the range of $100,000 to, say, $1 million.

insurance may just be the rationale the owners are looking for in order to close up, lock, stock, and barrel.

To continue to operate uninsured ("going bare") is very hazardous, even when regarded as a short-term solution. Nevertheless, many companies are doing just that while they seek a more acceptable solution to the insurance problem. Potential solutions include the following:

1. Self-insurance (complete or partial)
2. Trade association insurance pooling
3. Lobbying efforts

Complete self-insurance is really only feasible for the largest companies. Even these companies may want umbrella coverage as a hedge against catastrophe. Smaller companies may seek policies having high deductible features or clauses that obligate them to assume all or a portion of the legal defenses should such a need arise. It is doubtful if some of these smaller companies could withstand a number of substantial product liability judgments. Lack of comprehensive insurance coverage could mean that judgments become unenforceable, thus adversely affecting consumers. In the final analysis, self-insurance is probably synonymous with lack of insurance for the small company.

Of course, both manufacturers and insurers believe that the ultimate solution to the product liability problem lies in legislative reform. There is therefore a strong lobbying effort at both state and federal levels. Even if successful this will be a long and slow process, because it must be remembered that reform successfully brought about in Ohio has little impact in California, and the manufacturer is normally doing business in both states. This will be discussed in more detail below.

### The Insurance Industry

Product liability insurance is normally provided by an insurance carrier. In addition to providing insurance coverage, the insurance carrier can provide other services to its insured. Such services include expertise in loss control and in claims investigation and defense.

Insurers have broad experience and expertise in loss control. Working with the manufacturer, the insurer is often in a position to provide advice for controlling and minimizing product liability risk. Indeed, the insurance industry is currently placing economic pressure on the manufacturer to recognize the importance of products-loss control programs.

Insurers also offer considerable expertise in claims investigation and defense. Most manufacturers cannot afford to have defense personnel in-house. Consequently, the manufacturer should recognize the insurer's expertise in this area and use it when claims occur.

Despite such services, the insurance industry has suffered a loss of credibility in its handling of product liability insurance. This is due to clumsy rate making processes and past misjudgments on the part of insurance premium underwriters.

This image is partially due to the fact that product liability insurance is usually written along with other coverages as part of the comprehensive general liability. This has made it almost impossible to provide a separate and accurate profit-and-loss picture for the product liability losses; the rate-making process is not as actuarially precise as it would be in, say, life insurance. This leads to overly conservative rate making and high insurance costs. Clearly the insurance industry must discontinue this practice of blending all liability exposures, so that product liability can be rated as a separate item.

It would also be helpful to the insurance industry's image if product liability risks could be experience rated. This would eliminate the situation in which a small company is faced with a substantially increased premium when it has never had an adverse loss experience. Insurance companies contend, though, that past experience of small manufacturers may have little bearing on what might happen in the future. This is especially true for manufacturers of old products still out in the field that are not protected by a statute of limitations. Even though there is justification for the insurance industry's position, the practice of essentially ignoring a company's experience makes the company feel as if it is paying for somebody else's mistakes (including physicians' malpractice).

For an understanding of the intricacies of the insurance business, the engineer must also understand the concept of subrogation. This is the transfer from the insured to the insurer of rights that the insured may have in suing a third party. In other words, if the insurer of the victim accepts liability to indemnify the victim, the insurer then is subrogated to all rights of the victim against all other parties.

A good example of subrogation is provided by considering the case of a machinery-related accident occurring to an employee during the course of his or her employment. The employer is, of course, generally protected against liability by Workmens Compensation, and the Workmens Compensation insurer, whether private or state, will then have to make payments to the injured employee. The third party in this example could be the manufacturer of the machinery involved in the accident. If the injured employee brings suit against the machinery manufacturer and wins a judgment, the Workmens Compensation insurer will be reimbursed for payments already made. This is a classic products liability scenario in which one insurer (Workmens Compensation) is pitted against another (the machinery manufacturer's insurer); the Workmens Compensation insurer is said to be subrogated to rights of the insured employee. Many manufacturers believe that subrogation is responsible for spawning these third-party actions in which one insurance company sues to recover from another and thus contributes to the escalation of the product liability problem.

### The No-Fault Concept

The idea of no-fault compensation is that an injured person could recover damages for injury or other damages regardless of his or her fault or even misuse of the product. The public has been exposed to this idea in the area of automobile insurance.

Supporters of no-fault say it would provide prompt, efficient, and just compensation to the injured party while at the same time reducing insurance costs. Claims administration costs are now estimated to be 32 cents out of each premium dollar; since no-fault would reduce court costs, attorney's fees, and the size of awards, it is felt that this compensation system would result in considerable savings.* It would also free the courts from heavy case loads.

Although the manufacturer might end up paying a greater number of injury victims, each would likely be paid less because of the probable elimination of an allowance for "pain and suffering."

### Legislative Reform

There are many who believe that the liberal awards being granted by the courts and the contingency fee basis for compensating lawyers engaged in product liability actions have resulted in an unacceptable risk and cost of doing business. As indicated in the previous section on insurance, these risks have driven the insurance cost to approximately 1 percent of the sales dollar. In view of the fact that the profit in many industries ranges between 3 and 5 percent of the sales dollar, it is obvious that product liability costs are disproportionate.

The present situation can have other grave effects on our industrial system. Manufacturers are understandably hesitant to bring out new items or use new materials for which the chance for accident is unknown to top management. It is far more prudent to invest money in service industries, where the risk is smaller. Thus, we may see a shift away from the manufacture of hardware, in which high investment is involved.

Product liability also may have the effect of limiting competition. Insurance costs tend to be a larger part of the sales dollar for small businesses. Thus, as potential entrepreneurs study the feasibility of starting a business, the high cost of protecting themselves from financial ruin is one more factor to dissuade them. Some examples of awards that frighten manufacturers are the following:

- A judgment for $5.3 million against the manufacturer of a football helmet in favor of a player injured when a severe tackle forced the back edge of the helmet against his neck
- A judgment of $800,000 against a paper-making machine manufacturer for an injury that occurred on a 14-year-old machine that had been altered.

---

*Automobile no-fault experience, however, indicates that some of the projected savings do not occur.

In assessing potential remedies to the product liability problem, it is important to keep in mind the following interests:

1. Workplace versus nonworkplace
2. Short term versus long term
3. State versus federal
4. Consumer versus supplier

Underlying these interests is a basic social clash between those who view the tort-litigation system as a compensation system and those who view it as a system based on fault.

In view of what has happened to the insurance industry and manufacturers, it is probably safe to say that they would view tort law as a system that should be based on fault. Remedies advocated by them will therefore naturally reflect the feeling that tort law has gone too far in removing fault when assessing product liability. In other words, their position is that since the present problems are largely a result of legal changes made during the past 15 years, the remedies must also be legal. Both insurers and manufacturers have been acting to educate the public and lawmakers about the costs of product liability. They have also been pressing for the elimination of what they believe are unreasonable aspects of the existing situation. Major changes proposed include the following:

1. Enactment of reasonable statutes of limitations
2. Increased emphasis on safe use as well as on safe manufacture
3. Elimination or restriction of punitive damages
4. Assessment of defense costs against the plaintiff when suits are found to be without merit
5. Establishment of limitations on awards for noneconomic loss
6. Use of annuity-type awards
7. Use of arbitration to replace jury trials
8. Establishment of the no-fault concept
9. Modification of the present liability theory and a bolstering of the defense position
10. Establishment of limitations on contingency fees for plaintiff's attorneys

The above list of reforms can be viewed as representing interests of the supplier of products for the short term to provide some immediate relief.

The engineer is likely to be in accord with some of the manufacturers viewpoints; therefore, he will be able to support some of these proposals. For example, the engineer is not likely to accept the proposition that a manufacturer should retain liability risk for a product manufactured 40 years ago. This is because the engineer is aware of changes in state of the art, equipment misuse and modification, and improper maintenance or supervision. The engineer will therefore be likely to advocate some form of statute of limitations on liability risk.

Two types of statutes have been proposed. The first would have a statute of limitations that would apply only to the time running from the time of injury. Another proposal would have the statute run only from the time that the machine is first placed in the stream of commerce; this is called a statute of repose. Such statutes would presumably prevent situations such as the case in Illinois involving a machine sold in 1951 for $15,000 and resold years later as used equipment for a third time for $750. Most states do not have such statutes.

Most engineers would probably also support numbers 2, 3, and 4. Except in the most flagrant cases it seems inconsistent to make manufacturers subject to punitive damages while at the same time freeing the plaintiff from any responsibility for costs incurred by a defendant in defending against opportunistic, frivolous, or spurious allegations. Punitive damages, as distinguished from compensation for direct losses sustained by a victim of an accident, are intended to punish a defendant for gross misconduct and deter him from engaging in future misconduct.

Numbers 5 through 8 represent measures to deal with a spiraling record of awards, both in amount and probability. In 1973 there were two awards that approached $2 million each. In both of these cases, the awards greatly exceeded the economic loss suffered by the injured party. Consider the fact that at only 5 percent interest, these principal amounts would earn some $100,000 per year while maintaining the principal intact. Obviously, the juries making these awards considered "pain and suffering" as highly important. Number 5 proposes placing a ceiling on the portion of the award for pain and suffering. One proposal would limit the amount to $100,000 or some multiple of the economic loss, whichever is less.

Number 9, which proposes modifying present product liability theory to bolster the defense, implies another complete list of reforms. For example, the Policy Program Advisory Committee of the National Association of Manufacturers (NAM) approved an 8-point resolution on September 19, 1977, which called for the following:

1. Federal or state legislation should be enacted providing equitable treatment for manufacturers in product liability actions, but Workers Compensation should not be viewed as the exclusive remedy for workplace injuries.
2. A statute of repose commencing from the date on which a product enters the stream of commerce and ending on a date certain in the future should bar any recovery beyond that date from a manufacturer or seller.

3. The safety of a manufacturer's products should be judged by the state of the art at the time the product entered the stream of commerce.
4. The manufacturer of a product should not be held liable for alterations or changes made by others which substantially caused injury or damage.
5. If plaintiff's misuse of a product substantially caused or contributed to his injury, recovery should be barred.
6. The manufacturer's duty to warn of hazards should be judged against the actions of a reasonable person and his ability to identify such hazards.
7. Punitive damages should not be allowed in product liability actions.
8. Plaintiff's other sources of recovery should be admissible as evidence in product liability actions.

A national law that includes those points of the NAM resolution that are found to be valid would be an attractive solution to the problem. First, it would give manufacturers a clear indication of what is expected of them in providing a safe product. At present, what is considered an adequate warning label in Ohio may not be so judged in California. Do manufacturers with national distribution design to avoid liability for the state dealing most harshly with them?

A second advantage to a national law is that this would do more toward giving insurance companies what they need to establish insurance premiums than could ever be obtained by laws that vary from state to state. Imagine the difficulty in establishing insurance rates for a company in view of variations in the laws and variations in product distribution from state to state.

Finally, many participants in the product liability turmoil believe that the legal profession is abusing the contingency fee system. The United States is the only industrialized country with a contingency fee system. Under this system, the plaintiff's attorneys are permitted to collect fees that usually range from one-third to one-half of an award or settlement. Its original purpose was to provide legal counsel to those who would otherwise be unable to afford the cost of retaining a lawyer. This gives the lawyer a stake in the outcome of a case, and it is believed by many to generate lawsuits of little or no actual merit as a device for forcing a manufacturer or insurer into a settlement. Limitations on these fees would enable the injured party to receive a larger portion of the award and would probably place some restraints on lawyers who are aggressively seeking new areas of profit.

## Reference

1. Defense Research Institute, *Guide for Management, 2* (1972).

# 8

# Product Liability Litigation

Product liability litigation will almost inevitably revolve around some perceived deficiency in a product. If a deficiency exists, its ultimate source may lie in design negligence, in design defect, in production defect, or in a combination of these factors.

The issues of defect, of technical causation, and of design choices are seldom so simple that they will be easily disclosed. If the situation were obvious, the case would undoubtedly have been settled out of court. It is more likely that the technical facts are complex and esoteric. When these are coupled to the complexities of product liability law, a situation will usually exist in which reasonable people might disagree about where responsibility lies. This generally calls for a trial by jury.

In order for the product liability litigation to focus on the dominant technological questions, it is usually essential that engineers be involved in the litigation process as expert witnesses. It is the engineer, after all, who knows most about the product—its design, its manufacture, its cost, its utility, and its propensity to cause harm. The very nature of product liability law is such that lawyers, if left to themselves, would end up focusing on many irrelevant issues. This can be avoided if the legal and technical experts work together as a team.

### The Engineering Expert Witness

One definition of an expert is "a person with knowledge and technical ability in a given field, gained through education or experience, and the ability to articulate this knowledge, which makes this person more of an authority on the subject than the layman." A more pragmatic definition would be "a person qualified by a judge to act as an expert witness."

Almost every serious product liability suit involves engineering expert witnesses.

These engineering experts are needed to assess the evidence necessary to establish or refute the defect and causation issues. In the case of an alleged production defect, this will require identification of the following:[1]

1. The flaw or flaws relative to manufacturing or physical property standards
2. Evidence that the product's failure or malfunction is directly attributable to the flaw
3. The relationship of the failure or malfunction to the product's expected performance standards
4. The causal link between the failure or malfunction and the injury

For an alleged design defect, the engineering experts' assessments require the following:

1. Identification of the design error or errors that occasioned the injury
2. Enumeration of alternative design features proposed to reduce the danger
3. Evaluation of these features relative to the expected performance standards of the product, as well as their effect upon the product's subsequent usefulness and cost
4. Comparison of this product with other similar products
5 Establishment of the causal link between the design defect and the injury

Both plaintiff and defendant will generally need to employ at least one engineering expert capable of synthesizing these diverse elements to advise their respective attorneys. Later, if there is a trial, the experts will be expected to serve as witnesses and to give their opinions about defect and causation in terms and language that the jury and the judge can understand.

### *Sources and Qualifications of Expert Witnesses*

The determination and identification of the expertise necessary for qualification by an expert witness depends on the nature of the case. Possession of the requisite expertise is also a function of which party to the lawsuit the expert will assist—plaintiff or defendant.

The obvious source of expert witnesses for the defense of a product is the engineering staff of the manufacturer. Nobody knows more about a product than the engineers who designed and manufactured it. The project engineer who led the design team should be well qualified to defend it. The chief engineer of the company or someone of comparable stature also may act as an expert witness for the defense, but he will certainly rely heavily upon information provided by the project engineer.

Both defendant and plaintiff will sometimes make use of engineers who work for manufacturers of competitive products. The jury is likely to assign added credibility to testimony of an engineer of a competitor, especially if he testifies for the defense. However, participation in a trial usually involves such a commitment of time that many engineers with regular jobs decline to participate.

The defense attorneys will often employ an independent outside expert in addition to the company experts. An outside engineering consultant can often bring a broadened perspective and experience into product liability investigation and litigation. The witness stand is no place for a company engineer who is inarticulate, temperamental, or unprepared for rugged cross-examination.

The identification of a plaintiff's expert witness having experience with a specific product is usually a more difficult task. There is no open market for such experts, and plaintiffs are thus forced to seek generalists such as consulting engineers, academicians, and technicians. In very few instances will there be a precise matching of an expert's knowledge and the plaintiff's particular need for expertise. Consequently, the expert will have to be capable of undertaking self-education in the particular aspects of the product in question.

Regardless of whether it is for plaintiff or defendant, self-education by the expert is fair prey for a cross-examination suggesting that the expert lacks seasoned experience and expertise in the problem at hand. This is contorted logic, because the very nature of engineering is adaptability of expertise. While the particular function of components such as shafts, bearings, gears, switches, controls, hydraulics, and so forth may differ, the fundamental considerations in the design of these elements are the same; the basic principles of engineering do not change with time. If, therefore, the potential expert's credentials are impeccable and there are no special reasons for requiring seasoned experience over recently acquired understanding, then self-education should be recognized as a legitimate basis for qualifying an expert witness.

The ideal expert witness is one who has not only a broad background in scientific and engineering subjects but also considerable practical experience in the specific physical phenomena involved in the accident. If, for example, the case hinges on the rupture of a steel strut, the expert should be familiar with the appearance of that material when separated by sudden impact, fatigue, or any other means. Similarly, if an electric arc is suspected of being the origin of a fire, the expert should be able to recognize the characteristic appearance of metal melted by arcing as opposed to melting by fire. It is seldom an easy task to determine whether the damaged electrical conductor started the fire or the fire damaged the conductor. Again, self-education may be necessary. Engineers are accustomed to library research and to experimentation. It is appropriate to use both of these techniques in establishing the theory of the accident or in evaluating the theory of the opposition.

The expert witness needs to have the ability to recognize and carefully record the smallest detail about the product in question or at the accident scene. It is not enough to make a superficial appraisal. Trials are often delayed for a variety of reasons. Remembering details clearly enough to answer questions on the witness stand may be embarrassingly difficult by the time the trial is held.

It is important that the expert be a registered professional engineer. In most states, he cannot advertise himself as an engineer unless he is registered. It is a fine point as to whether or not calling oneself an engineer on the witness stand is advertising. The ensuing debate is best avoided. For example, in a federal court in Cincinnati in 1954, a manufacturer was sued after a woman was electrocuted by touching a fan guard while her kneee was against a hot water radiator. The chief engineer of the manufacturer took the stand and was subjected to a lengthy and upsetting period of questioning as to how he could use the title of chief engineer in view of his lack of an appropriate college degree and professional registration. He was visibly affected by the questions. The present climate is likely to be even more critical of any lack of credentials than it was in that more placid time of product liability.

A number of organizations specialize in supplying expert witnesses. The Defense Research Institute (DRI) has developed an Expert Witness Index of over 1900 names categorized according to field of expertise and geographic location. Resumes of potential witnesses are provided to attorneys at no cost.* Another organization, the Technical Advisory Service for Attorneys (TASA), also is a source of experts in a broad range of technical areas. When one of TASA's witnesses is retained, the billing is done through TASA, which adds a management fee to the witness's fee.†

### *The Relationship Between Expert and Attorney*

Suppose you have been contacted (usually by telephone) by an attorney regarding the possibility of your acting as an expert witness. The attorney may have obtained your name from a colleague, by word of mouth, or from one of the agencies discussed above. If you are a novice, you should know what you are getting into. You may be a fine engineer used to working broadly in an environment where people work collaboratively to achieve mutual goals. The courtroom is quite a different arena, in which the legal adversary system can reduce you to an ineffective witness. You must be prepared for a much narrower role and for tough cross-examination. You may have to disagree with equally competent engineers on the other side, and you absolutely must be prepared to make the time commitment necessary to do a thorough, complete investigation.

---

*For more information, or to register for the Expert Witness Index, contact DRI, 1100 West Wells Street, Milwaukee, Wis. 53233.

†For more information about TASA's services, write to TASA, 428 Pennsylvania Avenue, Fort Washington, Penn. 19034.

At the first meeting with the attorney, it is important to establish the proper professional relationship. The attorney will generally view himself as the primary director of the litigation, with you, the engineer, relegated to a secondary service position. Frequently the attorney will have a highly opinionated theory of the case and will want you simply to fill in the evidentiary gaps. He may assume that you know little about the legal criteria of product liability or that your knowledge of these legal matters will have little bearing on how you should proceed with your investigation.

The best working relationship is one in which the engineering expert becomes a resource and coequal partner rather than just a technician in the litigation. This requires that the expert not only serve the attorney as an evidentiary gap filler but that he also be knowledgeable about the legal criteria and how they impact on his or her investigation. One of the primary purposes of this book is to prepare the engineer for this professional role.

The nature of your first meeting with the attorney will also depend on whether he is representing the plaintiff or the defendant. If the attorney is representing the plaintiff, you probably will not be working in conjunction with company engineers also serving as experts. In this situation, the attorney will describe the general facts of the case. He may not know much about its techncial subtleties, and one of your major contributions may be to define, refine, and simplify the problems. However, you should never underestimate the ability, intelligence, or fighting nature of an attorney.

The plaintiff's attorney may ask you to investigate possible causes of the accident, to examine the evidence, and perhaps to examine the accident scene. You may be asked to evaluate the design of the product involved and to compare it to appropriate standards as well as to search for defects and possible failure of quality control. Usually this is done soon after the accident, because the plaintiff's attorney will want your results before preparing the formal complaint. You may be asked to prepare a formal report of your investigation, although the recent trend has been to deemphasize written reports.

After the complaint is served on the defendant, the so-called discovery process takes place. During this extensive phase of the proceedings, both sides will exchange information, take depositions of the opponent's witnesses, and perhaps make preliminary efforts to negotiate a settlement out of court. A deposition is a session separate from the trial in which the witness is questioned by each attorney, under oath, with a court reporter present. Be careful not to be lulled by the opposing attorney during a deposition session! A deposition is just like testimony; it is not simply informal conversation. Every word and question has great significance, and a careless answer may come back to haunt you later if there is a trial.

If the attorney is representing the defendant, your first meeting with the attorney is likely to occur much later in the time span between accident and trial. This is because the defendant's lawyer will have to wait until certain legal steps have been taken by the plaintiff. Also, the defendant's attorney may hope for a settlement out of court, which would allow him to avoid the time it takes to line up a full complement of witnesses, or he may want to defer expert witness charges by bringing them in at a later time.

At this first meeting the defense lawyer usually will provide his expert witness with a verbal description of the case as he understands it. He may ask questions about situations he fears are important. As an example, in a high-loss fire case the attorney defending the owner of the store in which the fire started was haunted by the thought that a circuit breaker with the handle taped in the "on" position (reported as fact) could not disconnect the circuit under current overload. He was much relieved when told that the circuit breaker carried the Underwriters Laboratories label and that their standard requires all such devices to trip free even if the handle is restrained.

The attorney will probably also provide the defense expert with a copy of the complaint. This formal legal document will usually contain a combination of two or more of the following theories of liability:

1. Misrepresentation in catalogues or advertisements, fraud, and deceit
2. Improper design of the product, lack of safety devices, or failure to produce a product in accord with state of the art
3. Negligent construction or assembly of the product, use of defective material, or inadequate quality control
4. Failure to warn of a defect or unobvious dangerous condition
5. Failure to repair the device after knowing of a defect
6. Breach of express warranty, either as a part of the sales contract or as a result of advertising
7. Breach of implied warranty or merchantibility or fitness for a specified use
8. Strict liability in tort

Before the trial the plaintiff may decide to abandon one or more of the theories given in the complaint because as the discovery process takes place, the original theories may become untenable.

The reports of the plaintiff's expert witnesses, depositions of the witnesses, and the physical evidence are usually in the hands of the defense attorney at the time of his first meeting with the defense expert, and this material will probably be entrusted to the expert for his study. The expert will be expected to respond based on his findings. The attorney may ask you to determine if the physical condition of the evidence is as reported by the opponents, if their conclusions are technically sound, and whether or not there are alternative technically sound scenarios of the accident.

The defense attorney undoubtedly hopes he can show where the plaintiff's experts erred; however, even if you conclude that the opponents are correct, this is valuable. In that event, the defense attorney will probably seek a second opinion. If others agree that the complaint is obviously correct, there is little point in going to trial to lose. Serious negotiation will undoubtedly begin at once.

Several other important things should occur at the first meeting. You should decide whether or not the case is one to which you feel qualified to make a contribution. Also, you should decide whether or not you and the attorney seem to communicate well. Although misunderstanding the way he puts questions in your office is of little consequence, it can be of overwhelming consequence when you are on the witness stand. He will be evaluating you in regard to these same points. He may also try to determine how you will react under the pressure of hard questions on the witness stand. Will you become rattled, angry, or flippant?

Finally, a very important thing to settle at the first meeting is your compensation. Bring the subject up if the attorney does not. In contrast to the contingency fee arrangement under which the plaintiff's attorney is compensated, it is considered unethical to base your compensation on the outcome of the trial. Also, the compensation is not for the expert's "opinion," because that could be interpreted by the jury as biasing that opinion. Your rate should be sufficient to ensure that you will make the necessary commitment in the case no matter what the pressure or time deadlines may be. It should be an hourly or daily rate and expressed as such if the opposing attorney wishes to question the financial arrangement when you are on the witness stand. At present, fees of $200 to $500 per day, plus expenses, are not unusual.

As the pretrial discovery process continues, keep the attorney informed of the time you spend. Tests and reports sometimes take more time than is understood by those without technical training. All involvement, including telephone conversations, should be counted. It is wise to keep a diary in case an accounting of your time spent is requested.

### *Investigation and Preparation*

Before beginning your investigation, you should insist that your attorney explain to you the legal theories both sides may use and the potential evidence or factual problems each will have. It is important to understand the issues to be litigated so that you won't waste time on things that will never be used in court. Make sure you understand what the legal words mean and what instructions are likely to be given to the jury.

After your attorney has done his homework and educated you, it is your job to do your homework and educate him. It is extremely important that you be thorough and methodical in your investigation and preparation.

Although it is desireable to be able to inspect the subject product and the accident scene immediately after the accident has occurred, in many cases it may be months or even years afterward that you are actually involved as an expert. In any event, the evidence must be made available to you, and if it comes into your possession, you must handle it carefully and responsibly.

In examining the evidence, make a permanent record of every important detail, using sketches and taking photographs. (Frequently you will have photographs made available to you by your attorney.) In examining the evidence you should keep in mind that you may never have another opportunity to examine it and that it may be months or years before the trial. When you finish examining the evidence, make a record of who you gave it to and obtain a receipt if you surrendered it to the opposing side.

In addition to the evidence, you should read all the documents, standards, and depositions pertaining to the case. Then you should meet with your attorney to discuss tests that may be required, exhibits, and analyses, all of which will involve costs. The expenses may be considerable in order to do a thorough, professional, investigation. If the costs are a problem, you may have to withdraw from the case. This is preferable to conducting a superficial investigation.

If tests are necessary, you will either have to conduct them yourself or arrange to have others do the testing under your supervision. Again, it is important to have complete documentation and photography.

During the course of your investigation you may want your attorney to elicit information from the other side, using either interrogatories or the discovery process. It is also important for you to imagine the scenario that may ultimately take place in the courtroom and to anticipate how the trial may unfold. Your mission will be to educate your attorney and perhaps ultimately the judge and jury also. Keep the theory of the case in mind and ask yourself the following questions: What was foreseeable? What were the design tradeoffs? What were the hazards? What were the alternative designs, safety devices, warnings, and standards? What is the defect, and is it unreasonably dangerous? What is the utility of the product, and what is its cost? How does it compare with competitive products, and what is the alternative to its use?

Frequently, the "possibility" and the "probability" of some event will arise. These words have special legal meanings quite distinct from their usual meanings in engineering. Something is probable under the law if it is more likely to occur than not. This requires a probability of only 51 percent or greater. Thus the engineer, accustomed to thinking something is probable only when its probability is appreciably higher than 51 percent, must be careful in discussing probability and certainty.

Whether you are working for the plaintiff or the defendant, the ultimate purpose of your examination should be viewed as the preparation of a report (which your attorney may or may not want) describing what you observed and stating your opinions. This report will be studied by your attorney, and if it is written it will probably become available to the opposition. This may prompt the opponent's attorney to ask for your deposition. In addition to simply gaining information, the deposition process will permit the opposing attorney to test your qualities as an expert. He may try to ask questions for which you are unprepared in order to determine if he has a worthy opponent to cross-examine in court.

Be careful during the deposition process, and do not respond too quickly. This will give your attorney time to object and you time to think. The best answers are short and direct. Also, never volunteer any information not directly asked for.

### The Trial

Only a small percentage of all product liability suits ever get to trial. Pretrial investigation by both sides often results in a decision by the plaintiff to drop the case or to settle out of court.

If a trial is to be held, the lawyer you are working with will hold additional conferences and discuss the latest developments as the trial approaches. There may be a new theory developed by the opponents that will send you scrambling for counterpoints. You will also want to review the questions your attorney wishes to put to you on the witness stand and to anticipate those of the cross-examiner.

Your attorney will give you the time and place of the trial. Unfortunately, you must be prepared to accept delays. The previous trial may have taken longer than anticipated, or the judge may be tied up with other legal matters.

You will be summoned to the courthouse sometime prior to the time you are scheduled to testify. Again, even though the court wants to proceed in an orderly manner, do not be surprised if you find yourself sitting idle for a day or two waiting for your turn to testify.

You should be neatly dressed and businesslike. Be extremely careful to whom you talk in the halls outside the courtoom; it is best to be courteous but not to enter into conversation with persons who might also be experts for the opposing side or even jurors.

The plaintiff will present his case first. His attorney will attempt to prove that one or more of the theories given in the complaint is valid. After he rests his case the defense attorney will usually move for dismissal on the basis that the plaintiff did not prove his case. If the judge can be convinced, he can order a directed verdict and dismiss the jury. The trial is over.

More often than not, the judge will allow the trial to continue. The defendant then presents his case, and after he is finished, he will again move for dismissal. However, if the judge believes that in view of both presentations, reasonable people might disagree on the verdict, he will rely on the jury for its decision.

During these proceedings you will have testified under oath as an expert witness. The procedure is that during direct examination by your attorney, you will identify yourself and state enough of your educational and practical engineering background to convince the judge and jury that you qualify as an expert. As such, you will have special privileges not accorded to other witnesses. You can state what you observed about the evidence you examined, and you can state your opinions. General witnesses can state only facts. The judge and jury will decide how credible your opinions are. The important thing to remember is that an honestly given opinion does not violate your oath, even if it proves to be invalid.

When stating your qualifications, be modest and unpretentious. The jury may be offended by a pedantic person all wrapped up in his own importance. During your stay on the witness stand, be polite and speak clearly. Remember that the use of technical jargon will only confuse or irritate both judge and jury. Explain concepts in simple terms, and use analogies if possible. Courtroom exhibits and demonstrations can be effective but dangerous. Be sure you have reviewed any such exhibits or demonstrations with your attorney before using them.

The direct examination will be followed by a cross-examination by the opponent's attorney. His questions must be limited to the same subjects covered in the direct examination. Thus, your attorney can limit the scope of your testimony by purposefully limiting his questions to you.

During the cross-examination, the opposing attorney will generally try to discredit your testimony by having you make obvious misstatements or contradictory statements. Some attorneys may try to shake or irritate you through insults or similar tactics. It is of the utmost importance that you handle this situation well. Some guidelines are presented below to prepare you for the tension that can accompany a demeaning cross-examination.

In general, you should remain calm and answer all questions thoughtfully. If the opposing attorney asks questions that contain glaring technical errors, point out that it is not possible to answer the question as stated. If he asks more than one question at a time, either tell him he has asked more than one question or ask which one he wants answered first. But be courteous and uncondescending!

Make a special effort to give simple explanations without sacrificing technical accuracy. As an example, in cases in which fire is suspected of having an electrical origin, the question may be asked why the protective fuses or circuit breakers did not prevent the fire. This can be satisfactorily answered by pointing out that the filament of an incandescent bulb is certainly hot enough to cause a fire, but that the fuse or circuit breaker does not operate every time the light is turned on. The electric bulb is designed to have a white-hot filament, but this same condition can also occur in an accident. The circuit breaker is designed to respond to the magnitude of the current, and it has no way of discriminating between the filament of a light bulb and a similar, but dangerous, phenomenon. In this way, the false idea that every electrical fire can be blamed on a fuse or circuit breaker can be dispelled.

Having the jurors believe you is just as important as having them understand you. Impressive credentials will help, but in the last analysis it is a matter of how you project expertise, honesty, and integrity. Also, you must be extremely careful to not become an advocate; your role is to be impartial. Keep within your own area of expertise, and never let an attorney lead you into areas in which you may have to guess, perhaps wrongly. Your role is first and foremost to educate the parties involved.

When the attorney asks you a question, it is generally best to look at the attorney when responding, unless your back is to the jury. But make sure the jury hears and understands your response. This may require speaking to the jury if the answer is an involved or complicated one. Also, don't forget that the judge is just as important as the jury; he controls the whole proceedings.

Don't exaggerate, guess, hedge, or bias your answers to questions. Never hesitate to tell the truth, even though it may be detrimental to your attorney's client. If you do not remember some detail about the evidence, say so, and if it becomes necessary to change something you said previously, then do it. Remember, you are acting as an independent professional, which requires never prostituting your professional integrity.

### *Responding to Cross-Examination*

On cross-examination some attorneys may try to insult the expert, irritate him and make him angry, or otherwise discredit him. It will be very helpful to you if you realize that this process has nothing to do with you personally. It may even happen that after the case is over, the same attorney who badgered you and shouted at you will congratulate you for the job you did. Also, he may want you to work with him at some time in the future.

It is best to take a somewhat relaxed and detached viewpoint of the entire proceedings when on the stand. The following guidelines are presented to assist you in acting as a salesman for your profession.

1. Never lose control of your emotions or react to derogatory remarks, verbal abuse, or similar tactics. If you get into a shouting match, you are sure to come out last. You can gain credibility while the attorney loses credibility by being courteous, patient, and in control of your wits. After all, you are not a stooge, and generally the court is there to protect you from such browbeating. Always try to be cooperative, and remember that your role is one of being an impartial educator. The cross-examiner may ask you how much you are getting paid for your testimony, implying that your interests are purely commercial. He may also imply that you often testify for the side you are representing, so as to suggest a natural predisposition or bias for either defense or plaintiff. Just roll with the punches and maintain a sense of humor. But don't be flippant.

2. Do not answer questions you do not understand or guess at answers you do not know. Ask for clarification or ask gracefully that the question be repeated. If you don't know an answer, your best response is, "I don't know." Don't be too proud to admit that your expertise is limited.
3. Do not volunteer information. Listen to the question carefully, determine its basic thrust, and answer it concisely and simply. Don't get off on a tangent or begin to qualify all your answers. On the other hand, as an expert witness, you are not limited to just "yes" or "no" answers.
4. Be a salesman for your profession by being candid, objective, respectful, and helpful. You are providing a service in attempting to render justice in a civil dispute, and this is a very worthwhile and important one. Remember that all parties concerned are intelligent people anxious to resolve conscientiously a difficult task. Be a part of the solution and not of the problem.

### After the Trial

Most cases never come to trial but are instead settled by negotiation. This is usually done on the basis of the facts uncovered during the investigation and discovery phases, the theories evolved that seem probably to explain the accident, and the risk both attorneys perceive of losing the case. The opinions, research, and advice of the technical experts are just as important to the results when the settlement is negotiated as they are when the judgment is rendered by a jury.

If the case does go to the jury, there is little to do but await the verdict. Remember though, that even after you have testified, you should not discuss the case with the opposition. (Indeed, it is good practice to be discreet at all times.)

You are obligated to retain all files, exhibits, and data used in your testimony until the case is settled. And sometimes cases are reopened or appealed.

After the trial, you will certainly be interested in the outcome. Call your attorney and ask for this information, and also ask his opinion of your effectiveness as an expert witness. This is the only way to become a better witness in the future.

The litigation process can present interesting, demanding, and rewarding professional work for the prospective technical expert. By becoming involved, you will find many shortcomings and deficiencies in the process. Nevertheless, the system is a fair one and one in which the interests of justice are served well.

### Reference

1. A. S. Weinstein et al., *ASME Paper 74-WA/SAF-3* (1974).

# 9

# Engineering and Ethics

Prior to the Second World War, the ethical considerations facing the engineering profession were associated with problems of a rather parochial nature. A good example is the problem of injury to persons involved with power generation using fossil fuel steam-generating plants. Through the voluntary initiatives and ethical commitment of individual engineers, the American Society of Mechanical Engineers (ASME) developed their Boiler Code, which produced a notable record of safety. The resulting image of the engineering profession became one of professional responsibility in designing and building such steam plants with full recognition of the obligation to protect the public health and safety.

Technological developments since the Second World War have been truly revolutionary and have resulted in the confrontation of the engineering community with social issues such as automation, energy, environmentalism, consumerism, and product safety litigation. The ethical dilemmas these issues pose have resulted in a rapidly developing tarnish on the promise of technology and on the image of the engineer. Whereas the engineer was once faced with ethical problems of relatively limited scope, now they seem diffuse and shrouded in complex processes within governmental and corporate structures.

Paralleling this loss in technical confidence is an increasingly sensitized society and a legal system that has shifted the burden of responsibility for safety from the consumer to the manufacturer. The result is a virtual explosion in product liability litigation.

The atmosphere in which product liability litigation occurs can be unsettling for the engineer because of the adversary nature of the proceedings and because a greater commitment is made to victory than to truth. Whereas the engineer is normally applying an absolute point of view, the function of the attorney is to advocate, within legal limits, his client's point of view—by first presenting his client's evidence in the best possible light and then weakening his opponent's evidence in whatever ways seem appropriate. Also, the basis upon which the engineering expert witness must make his decisions in evaluating hazards, the probabilities of risk, and the appropriate design response cannot be objectively quantified. These procedures do not conform to the analytic training of the engineer which fosters a desire for well-defined rules.

Ethical problems are, indeed, most difficult for the engineer to fit into his pattern of thinking. There are no analytical models for assessing and solving them. As an expert witness, for example, the engineer will more often than not be required to balance subjectively factual evidence, statements of other witnesses, esoteric knowledge of design or production considerations, and economic and psychological concerns. The less quantifiable the data, the more subjective must be the conclusions. Yet the conclusion must be made!

One result of the present emphasis on product liability is to define more clearly the *total* responsibility of the engineer. His primary duty is, and has been, to his employer, but it is not exclusively there. The engineer also has a responsibility to the consumer and to his profession. Persons engaged in learned occupations have long recognized that their education and training carries ethical responsibilities and that they have coordinate obligations to their employer and to the consumer.

Although there is no substitute for individual action based on a firm philosophical and ethical foundation, engineers have developed guidelines for professional conduct based on the experiences of those many engineers who have had previously to wrestle with troublesome ethical questions and situations. These are contained in the *Codes of Ethics of Engineers.*

New engineering graduates are often totally unprepared for problems or situations that will test their professionalism and ethics. Indeed, the engineer is often unaware that an ethical problem exists. Furthermore, codes of ethics are frequently of little value in the many gray-shaded ethical situations that may arise. Consequently, the engineer must cultivate an awareness of the importance of being able to create, from a solid philosophical background, his own rules of ethical behavior for guidance in these situations. This can be done by recognizing that ethics and ethical theories must, and can, be learned, just as engineering theories are learned. The basic difference is that engineering theories are based on actual, empirical behavior, while ethical theories are based on models of ideal behavior.

It must also be remembered that the autonomy of a professional person depends on how he is perceived by the public. This means that the engineer must not only act ethically, but he must also pay great attention to the appearance of his actions. An action that appears to be unethical should not be undertaken, regardless of the true ethical situation.

Codes of ethics provide guidelines for a number of recurring and likely kinds of ethical problems. Because codes of ethics are superior to, and go beyond, legal requirements, adherence to these codes represents a voluntary commitment. Nevertheless, the basis of any profession is its ethical underpinning, and every professional must either accept existing ethical guidelines or work actively to have them changed.

### Canons of Ethics

Whether reference is made to a code of ethics, rules of professional conduct, or a canon of ethics for engineers, the focus is on those principles of conduct governing an individual or a profession.[1] While the principles that make up a code of ethics remain fixed, their interpretations must be continually reviewed in the context of human and institutional interactions.

A number of professional organizations have published what they believe to be fundamental principles of engineering ethics. Most are similar to the following statement, which has been adopted by the Tennessee State Board of Architects and Engineering Examiners:

> The architect or engineer shall be completely objective and truthful in all professional reports, statements, or testimony. He shall include all relevant and pertinent information in such reports, statements or testimony.
>
> The architect or engineer when serving as an expert or technical witness before any court, commission, or other tribunal, shall express an opinion only when it is founded upon adequate knowledge of the facts in issue, upon a background of technical competence in the subject matter, and upon honest conviction of the accuracy and propriety of his testimony.

The Engineers Council for Professional Development (ECPD), which represents the 12 major engineering societies, has adopted a Code of Ethics of Engineers, which is reproduced in Fig. 9.1. Note that its fundamental canons address the engineer's relationship with fellow engineers, his employer, and the public. The recognition of these relationships preceded the product liability crisis by many years, and the recent legal activity merely proves that engineers have indeed addressed the need for guidelines governing their own professional activity.

### Ethical Considerations in Product Safety

The primary focus of our discussion of ethics has been on the safety of a product. When the legitimate use of a product, its foreseeable misuse and abuse, and all human error in using it are considered, it may become difficult to find any products to be totally safe. Nevertheless, it is also true that ways can usually be found to lessen the probability of some conceivable injury. However, safety

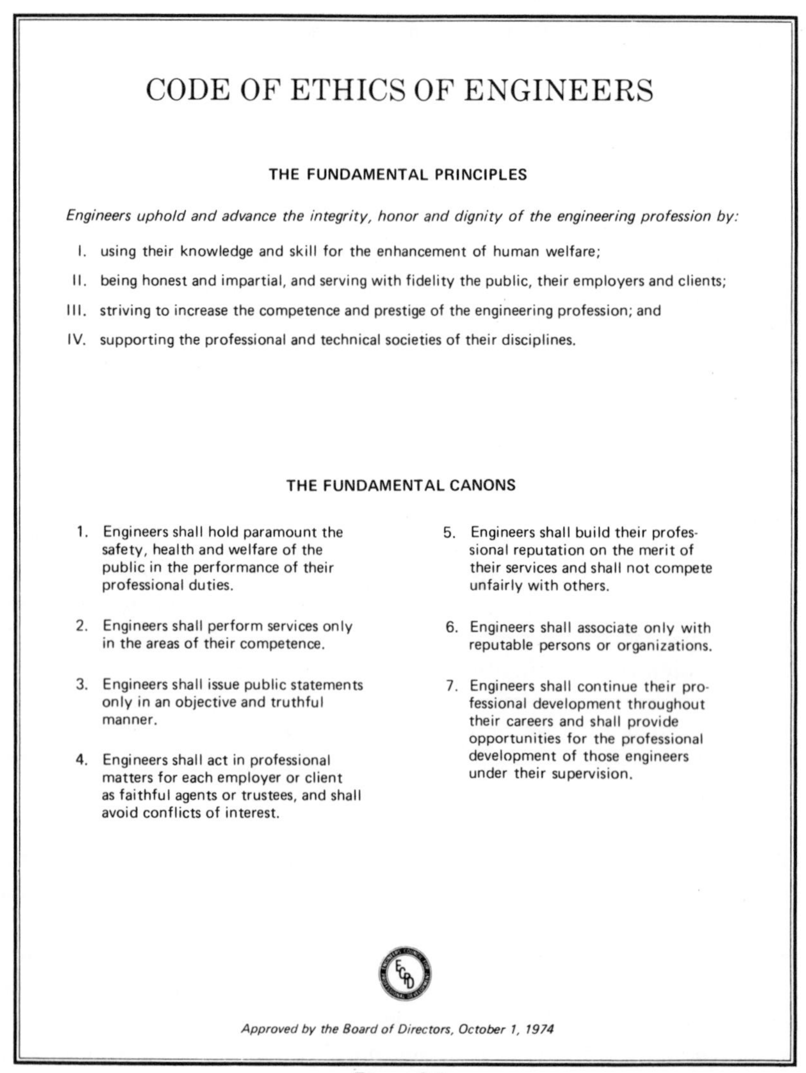

CODE OF ETHICS OF ENGINEERS

THE FUNDAMENTAL PRINCIPLES

*Engineers uphold and advance the integrity, honor and dignity of the engineering profession by:*

I. using their knowledge and skill for the enhancement of human welfare;

II. being honest and impartial, and serving with fidelity the public, their employers and clients;

III. striving to increase the competence and prestige of the engineering profession; and

IV. supporting the professional and technical societies of their disciplines.

THE FUNDAMENTAL CANONS

1. Engineers shall hold paramount the safety, health and welfare of the public in the performance of their professional duties.

2. Engineers shall perform services only in the areas of their competence.

3. Engineers shall issue public statements only in an objective and truthful manner.

4. Engineers shall act in professional matters for each employer or client as faithful agents or trustees, and shall avoid conflicts of interest.

5. Engineers shall build their professional reputation on the merit of their services and shall not compete unfairly with others.

6. Engineers shall associate only with reputable persons or organizations.

7. Engineers shall continue their professional development throughout their careers and shall provide opportunities for the professional development of those engineers under their supervision.

*Approved by the Board of Directors, October 1, 1974*

**Figure 9.1**

usually costs money, and a supersafe product that must be sold at a price above that acceptable to consumers does not sell and thus does not prevent accidents. It may merely drive consumers to a less expensive and less safe product of a competitor.

It is also true that manufacturers must be able to sell their products and make a profit in order to survive. Even the courts and laws do not require that a manufacturer price his products to the point of extinction. Thus, the problem of balancing risk against utility, discussed earlier, is ever-present.

It is the engineer who finds himself squarely in the middle of the problem of risk and utility. On the one hand, it is the engineer's duty to design a product that gives a fair, if not maximum, rate of return on investment to his employer and on the other hand, as stated in the Code of Ethics, it is his duty to "hold paramount the safety, health, and welfare of the public." To some extent, these are competing interests. Of course, with product liability now a significant part of the cost of doing business, a mandate has been given to the engineer to be more concerned about the safety of the product than in the past. Indeed, to not give high priority to the safety of the product is unethical.

A most difficult situation occurs when an engineer or group of engineers believe that a product or system is defective—either in design or production—such as to make it inherently dangerous. Holding paramount the safety, health, and welfare of the public, the engineer has the duty to inform his employer if it is his judgment that the product is unsafe. Judgments are based on analyses—either formal or heuristic—and analyses usually contain assumptions. Honest differences can exist over the reasonableness of the assumptions and the accuracy of the analysis showing the product to be unsafe. What, then, should be done?

There have been a few well-publicized cases recently in which engineers have publicly accused their employers of proceeding with the development of unsafe products or projects. The ethical problems arising in the aircraft brake debacle[2] and in the control system for San Francisco's BART system[3] are two examples in which engineers were faced with this difficult decision.

To take such an extreme action must be viewed as a last resort. Also, any engineer taking this action acts unethically if his accusations are motivated by the desire for notoriety or to be a hero or martyr. The engineer's motivation must be to protect the public's safety, health, or welfare from an employer who refuses to reduce significant hazards.

Unfortunately, the judgment on "blowing the whistle" usually cannot be made until many years after the act. The consequences of the individual's decisions are serious both to himself and to those who might be harmed if alleged defects are tolerated. These are most difficult decisions to make, and they require both a sound philosophical base and the courage to act ethically.

### The Ethical Engineer

In the musical *Camelot,* Lancelot sings that a knight of the Round Table should be invincible, succeed where a lesser man would fail, slay a dragon in record time, and swim a moat in a suit of iron mail. He goes on to suggest that even a miracle or two should not be beyond the knight's potential. Fortunately, engineers need not do all these things, but a description of what they should do is perhaps appropriate.

The ideal engineer is best described by the ECPD Code of Ethics, although perhaps with different emphases. First, the engineer should increase his own competence and prestige. He should become expert in the technical aspects of his profession; this will require continuing education. Second, the engineer should use knowledge and skill for the advancement of human welfare; it is a deep satisfaction to know that the products he develops helps others to have a better life. The engineer's contribution to human welfare goes beyond product design and includes many different voluntary activities, both on and off the job. Finally, the engineer must be honest and impartial and serve his employer, clients, and the public with devotion.

### References

1. R. J. Schemel, *ASEE Paper 1320,* ASEE Annual Conference (1976).
2. K. W. Vandivier, *ASME Paper 75-TS-2* (1975).
3. R. W. Anderson, *ASME Paper 75-TS-4* (1975).

# 10

# The Role of Engineering Education

The product liability crisis has an impact on engineering education through the teaching of design courses and, increasingly, through faculty consulting work. Furthermore, the risk of personal liability for the design engineer and the probability of confrontation with troublesome ethical questions is greater today than in the past. It is therefore incumbent on engineering education to provide the engineering student with an awareness of these potential pitfalls.

In the context of the social forces that are functioning today—new regulatory agencies, a broader legal framework, and a heightened social awareness—traditional design education, and indeed engineering education in general, must be reevaluated from the point of view of including considerations of product safety and other topics needed in a new era of engineering.

During the past quarter-century, engineering education has responded to the challenge of increasing the scientific content of its programs. This direction resulted from the experience during World War II that persons trained in science rather than engineering were required to develop weapons technology based on research at the forefront of science. As a result of that experience, the scientific content of engineering curricula was increased to train engineers capable of translating discoveries at the forefront of science into innovative products.

Paralleling and complementing these changes in educational philosophy, research efforts by engineering faculty members increased rapidly as government-sponsored research funds became readily available. This encouraged university administrators to recruit engineering faculty who could demonstrate research capability as opposed to the ability to design or to teach design. The doctorate became the educational requirement for faculty appointment, promotion, and tenure, and the number of schools offering graduate programs increased rapidly.

While these developments brought many needed changes into engineering education, they tended to emphasize analysis as opposed to synthesis and to produce analysis-oriented graduates. There was also a decrease in emphasis on laboratory experience in the curricula. Thus, competence in hands-on experimentation, which is such a necessary part of developing successful products, diminished. In short, the emphasis and resources of undergraduate engineering education were redirected in such a way as to produce graduates less qualified to design and develop products that are reliable and safe. This represents a submergence of traditions belonging uniquely to engineering. It is now time to review engineering curricula in the context of present-day societal demands.

### Product Liability and Design Engineering Education

In a new era of engineering, engineering education must accept the responsibility to formulate curricula that teach the subjects and skills needed by the engineer to invent, develop, and design products having the superior reliability and safety that consumers demand.

Foremost among the technical competences necessary to design safer products is the knowledge of human factors and human behavioral variables. Human factors information has been used extensively in designing military systems and hardware, when the operator may be placed in an unusually hazardous environment or the consequences of an error may be catastrophic. Design engineering education must develop an increased awareness of the importance of these factors for civilian products. Human factors specialists such as psychologists, physicians, and anthropologists can provide important design information to the engineer.

The properties of materials must also be introduced to the depth required to provide an understanding of how the environment influences these properties. The properties of alloys, heat treatments, and composite materials should be known to the extent that engineers will examine their literature closely before using them. Strange as it may seem, changes in material characteristics at high stress levels are often overlooked. Aging, oxidation of metals, work hardening, annealing, anisotropic effects, corrosion, and fatigue are often ignored in making design decisions. More modern courses are needed particularly in the area of failure and fatigue. Also, the impingement of foreign materials and of energy can cause degradation. This must be taken into account if the product is to be safe throughout the entire product life cycle.

The topics of reliability, maintainability, statistics, random processes, and hazard analysis should also be included in a curriculum emphasizing design along with information on the design of experiments. These topics are extremely important in the mass production techniques that produce our products and standard of living.

Today the designer must not only be able to apply those technical elements discussed above but must also design to meet the expectations of society. This means using far broader design criteria than in the past, including such things as have been presented in this book. In particular, as part of the design process the criterion of "adequacy for user and use environment" must be addressed, together with engineering functional requirements. Some of the key elements that must be included are the following:

1. Identification of product use environments that encompass foreseeable misuses and expected uses, as well as intended uses.
2. Identification of the modes of operation within the use environments. These should include such considerations as shipping, storage, installation, starting and stopping, cleaning, repair, maintenance, salvaging, and disposal.
3. Systematic identification of hazards and risks associated with the use and modes of operation.
4. Failure or malfunction analysis of a design to identify other possible hazards.
5. Systematic identification of applicable standards, regulations, and all relevant data on product-caused injuries to aid in developing safety considerations.

In addition to these pragmatic measures, a new basis for decision making and a new design philosophy must become as intrinsic to design and design education as are those traditional technical considerations such as stress analysis, kinematics, and electrical circuitry. These new bases in design are those of risk-utility balancing, elements of design procedure for producing the reasonably safe product, and ethical considerations. Peters [1] and Peters et al. [2] have already discussed these concepts. In Ref. 1 a rather comprehensive treatment is given of design considerations that should be included in project design courses insofar as product liability prevention is concerned.

The thesis advanced in Ref. 2 is that the same risk and utility indicia should be used prospectively in design education (and practice) as are used retrospectively by the courts in attempting to distinguish between reasonable safety and unreasonable danger. In other words, since the decision-making process during litigation involves a full appreciation of the complex tradeoffs involved in product design, the design process itself (and design education) should reflect these same kind of considerations. The specific balancing factors to be considered by the designer in making risk-utility analyses are presented in the Appendix.

### Guidelines for Professional Practice

The time available in the curriculum for design education and the design process itself impose restrictions on the scope of topics that can be discussed. Yet product liability is representative of a whole new set of issues with which future engineers must be concerned. These broader questions relate to the professional practice of engineers.

In restoring the traditional professionalism and broad perspective of engineering education and practice, it is necessary to redefine the desirable competencies and learning activities that make up a professionally oriented curriculum. One essential competency is in the area of professional practice. Some of the topics that should be included are listed below.

### *Engineering and the Law*

The areas that should be covered are the following:

1. Elements of tort law, with the emphasis on definitions, negligence, breach of warranty, strict liability, and landmark decisions leading to the current product liability climate.
2. Guidelines for personal behavior in interacting with lawyers, in expert witnessing, and in forensic engineering. The engineer must be taught to see the lawyer's viewpoint (and the lawyer should be taught to appreciate the engineer's viewpoint).
3. Elements of patent law and of other intellectual property.

### *Engineering and Government*

Areas that should be covered are the following:

1. The complex nature and interaction of questions such as nuclear energy, nuclear nonproliferation, and military engineering.
2. Federal and state regulatory agencies, such as OSHA, CPSC, FDA, and FTC; also, Workmens Compensation should be related to product liability.

### *Professional Practice*

Areas that should be covered are the following:

1. Professional registration, certification, continuing education, guidelines for employment, and an awareness of the limited nature of formal education; how the engineer must accept self-education as a way of life for the professional.
2. How to conduct an investigation, report preparation, and documentation; elements of expert witnessing.

3. The necessity of constant monitoring of legal actions affecting both the engineer personally (personal liability, for example) and his employer (or client).
4. A study of ethics, codes of ethics, and selected ethical and moral dilemmas.

A course covering these topics could be introduced early in the curriculum, preferably during the freshman year, when students are receiving a heavy dose of nonengineering subjects.

## Summary

The issue of product liability is raised when people are hurt or their property damaged allegedly due to product defect. To minimize product liability risk in designing products, the engineering graduate must be prepared to understand people, to know materials, and to know how to carry out both analytical and experimental evaluations to assess the potential harm of a product as well as its utility. This knowledge must be applied not only in the context of the intended use of the product but in every foreseeable use and misuse environment that can occur during the total life of the product.

Present engineering curricula do not sufficiently emphasize safety in product design. Most of the emphasis is on basic science and analytical methods. While these disciplines are necessary in an engineer's training, they are not sufficient. We must be careful not to forget to provide educational experiences that will give students the knowledge and incentive to do a careful, thorough job in design and design management. Engineering education needs to address this problem and to develop a program in which those students interested in design careers can learn how to be skillful and responsive to modern design requirements.

The engineering graduate today possesses superior analytical ability, has access to modern computational tools, and can eliminate much of the uncertainty that accompanied cut-and-try methods of the past. Nevertheless, engineering will always involve the art of synthesis and decision making in the absence of complete information. Professional engineering adds the element of responsiveness to human needs.

Engineering education can make a valuable contribution to society by recognizing that not all engineering students are suited by disposition for research and development, just as not all are suited for product design. As in the practice of engineering, the education of engineers needs to have balance, flexibility, and opportunities for professionals with diverse interests.

## References

1. L. C. Peters, *ASEE Paper,* in 1977 *Frontiers in Education Conference Proceedings.*
2. L. C. Peters et al., *Mechanical Engineering, 99* (1977).

# Appendix
# The Unreasonably Dangerous Product

The words *reasonable* and *unreasonable* pervade the literature of product liability. For example, the standard for negligence is what a "reasonable" person would or would not have done under the circumstances. Thus, the tort of negligence is closely identified with the so-called reasonable person standard.

The word *unreasonable* is also used in connection with the doctrine of strict liability. One who sells any product in a defective condition "unreasonably dangerous" to the user or consumer or to his or her property is subject to liability (strict liability). Thus, strict liability is closely identified with the concept of "unreasonable danger."

Other expressions used in the literature of product liability also abound in the use of the word *unreasonable.* For example, the Consumer Product Safety Act (CPSA) was passed to protect the public against "unreasonable risk" of injury associated with consumer products. Also, the expression "unreasonable hazard" is frequently used along with "imminent hazard" and "substantial hazard."

These many different expressions are often indiscriminately used to describe the same thing. This is a source of misunderstanding and misinterpretation, particularly when this lack of discrimination is added to an already nonexistent concrete and concise legal definition for what is meant by "unreasonable."

In the interests of precision, we would like to define a hazard as a potential to do harm, a risk as a probability, and a danger as a combination of hazard plus risk. Nevertheless, it is generally true that "unreasonable hazard," "unreasonable risk," and "unreasonable danger" are used synonymously in the bulk of the product liability literature. We will accept this. Yet there still remains an uncertain relationship between the meanings of "reasonable person" and "unreasonable danger."

The "reasonable person" whose behavior is made the standard is an imaginary person, vague and elusive. The standard is purposely a variable one, because it is not possible specifically to define in advance what a reasonable person would do in any individual case. The reasonable person is not to be identified as an ordinary individual, nor is he a juror or an average of what jurors would do. Yet, he is a prudent and careful person and the personification of reasonable behavior as determined by the jury's social judgment. He is, in the words of one legal commentator, "the excellent but odious character standing like a monument in our Courts of Justice, vainly appealing to his fellow citizens to order their lives after his own example."

It is also not possible to set down a standard, quantized definition to determine in a concrete way what constitutes an "unreasonably dangerous" product. To engineers and manufacturers used to having clear definitions and rules for operation, this vagueness is frustrating. It has also proved to be difficult for the courts.

In order to overcome this, attempts have been made to define the "unreasonable danger" concept of strict liability in terms of the "reasonable person" standard (which goes back many years in the common law). For example, in *Welch* v. *Outboard Marine Corp.* (1973), the court instructed the jury that "a condition is *unreasonably dangerous* so as to constitute a defective condition when it is so dangerous that a *reasonable man* would not sell the product if he knew of the risk involved." There are many such cases similar to this in which the courts have attempted to define unreasonable danger in terms of placing the manufacturer and vendor, or the consumer, in the role of the reasonable person. An excellent review of some of these cases, along with a comprehensive discussion of the difficulties in defining "unreasonable danger," are presented in Ref. 1 and Ref. 2.

Attempts to define the unreasonable danger criterion of strict liability in terms of the reasonable person standard is said, in an expression used by the courts, to "ring of negligence," because the reasonable person standard stems from the concept of negligence. Since the doctrine of strict liability in tort specifically excludes negligence as a criterion for manufacturer's liability, attempts to define the strict liability question in terms of the reasonable person represent an inconsistency.

In Ref. 1 it is pointed out that the strict liability question should not be defined from the perspective of the reasonable person but rather from the perspective of society making the ultimate judgment by balancing risk and utility. The unreasonable danger question, then, is posed in terms of whether, given its risks and benefits and its possible alternatives, we as a society will live with the product in its existing state or will require an altered, less dangerous form. The basic question is whether the product is a reasonable one given the reality of its use in contemporary society.

What are the balancing factors of risk versus utility in determining if a product is unreasonably dangerous? They are, according to Ref. 2, the following:

1. The usefulness of the product
2. The availability of safer replacement products
3. The likelihood of injury
4. The probable seriousness of injuries
5. The obviousness of danger involved
6. The public's expectations concerning the product's performance
7. The possible elimination of risk through care or warnings
8. The possible elimination of risk without impairing the product
9. The "state of the art" of the industry
10. The cost of making the product safer
11. Consumer willingness to pay for a higher-priced but safer product
12. The bargaining power of the manufacturer as contrasted with that of the consumer

The evidence necessary to address these balancing factors is what should be exposed by the parties to a strict liability action. The focus will be on the product completely described and judged based on its total milieu of use rather than on the manufacturer's conduct. Then the jury can subjectively integrate both the benefits and costs of the alleged "unreasonably dangerous" product to determine if it offends society.

### References

1. W. A. Donaher et al., *Texas Law Review, 52* (1974).
2. S. D. Hoffman, *Underwriters Laboratories/LAB DATA* (1975).

# Index